MIND OF THE MATTER, MATTER OF THE MIND

Paperback Version

Jensine Andresen

CONTENTS

INTRODUCTION

This book considers the academic and societal implications of *creative acculturation* with the advanced, extraterrestrial intelligence (ETI) operating many unidentified aerial phenomena (UAP) on and around Earth.

To be clear, I do not think that all UAP are extraterrestrial in origin. Numerous U.S. Government (USG) Special Access Programs (SAPs) have the technological competence to design, build, and fly drones and other aerial vehicles that may appear so unfamiliar to people that they are mistaken as extraterrestrial in origin. It also is likely that similar programs exist in other countries. For example, an object demonstrating plasma stealth technology could appear quite unfamiliar to human witnesses while nevertheless being an example of human technology (Rogoway and Trevithick 2019).

Nevertheless, my view is that many UAP are in fact extraterrestrial in origin. Considerable historical evidence and analysis supports my position (COBEPS 2021; McDonald 1969; Sturrock 1987; 1999; 2004; Sturrock et al. 1998). In addition, tens of millions of people have witnessed UAP, which also have been documented in photographs, on video, and with radar. My view that many UAP are in fact extraterrestrial in origin is supported further by plausible interpretations of various petroglyphs, geoglyphs, and artifacts, and by accounts described in the literature of many cultures extending back thousands of years.

I use the acronym UAP in a broad sense that includes what previously were referred to as unidentified flying objects (UFOs) and, also, unidentified submerged objects (USOs). My usage of UAP is consistent with how the term is defined in the U.S. Government's (USG's) National Defense Authorization

Act for Fiscal Year 2022 (NDAA FY2022, henceforth NDAA). The NDAA addresses UAP in Title XVI, SPACE ACTIVITIES, STRATEGIC PROGRAMS, AND INTELLIGENCE MATTERS, Subtitle E—Other Matters, Section 1683, ESTABLISHMENT OF OFFICE, ORGANIZATIONAL STRUCTURE, AND AUTHORITIES TO ADDRESS UNIDENTIFIED AERIAL PHENOMENA (NDAA 2021, 531- 33, 571-81). Section 1683(l), DEFINITIONS, states:

> (4) The term "transmedium objects or devices" means objects or devices that are observed to transition between space and the atmosphere, or between the atmosphere and bodies of water, that are not immediately identifiable.
>
> (5) The term "unidentified aerial phenomena" means—
>
>> (A) airborne objects that are not immediately identifiable;
>>
>> (B) transmedium objects or devices; and
>>
>> (C) submerged objects or devices that are not immediately identifiable and that display behavior or performance characteristics suggesting that the objects or devices may be related to the objects or devices described in sub-paragraph (A) or (B) (NDAA 2021, 583).

The NDAA's definition of UAP is slightly broader than the definition of UAP found in the unclassified portion of the June 25, 2021 document entitled *Preliminary Assessment: Unidentified Aerial Phenomena* issued by the Office of the Director of National Intelligence (ODNI). There, UAP is defined as follows:

> **Unidentified Aerial Phenomena (UAP):** Airborne objects not immediately identifiable. The acronym UAP represents the broadest category of airborne objects reviewed for analysis (ODNI 2021, 8).

In this book, my usage of UAP is consistent with the comparatively broader definition stated in the NDAA rather than the relatively narrower definition stated in ODNI's *Preliminary Assessment*.

CHAPTER 1

The U.S. Government's (USG's)-Eye View

Official USG interest in UAP goes back decades. Rear Admiral Roscoe H. Hillenkoetter, the Director of Central Intelligence (DCI) from May 1, 1947 to October 7, 1950 (CNN 2021), possessed a keen interest in UAP (Aerial Phenomena 2022). In addition, former President Harry S. Truman, who served from April 12, 1945 to January 20, 1953, was asked on camera by a reporter about UAP, "Did the Joint Chiefs of Staff talk to you or concern you about the unknown or the unidentified flying objects? I'm speaking of the flying saucers." Truman replied, "Oh yes, we discussed it at every conference that we had with the military and they never had [*sic*] been— never were able to make me a concrete report on it" (UFOs 2019).

From 1956 to 1980, the National Investigations Committee on Aerial Phenomena (NICAP), with which Hillenkoetter was affiliated, raised two concerns relating to UAP. The first concern was that if UAP formations were mistaken for Soviet missiles, this erroneously could be interpreted as a surprise attack by the former Soviet Union (the Union of Soviet Socialist Republics, henceforth USSR), accidentally triggering a war. The second concern was that the USSR could employ psychological warfare tactics by claiming that UAP were secret Soviet weapons against which U.S. defenses were ineffective (Aerial Phenomena 2022).

In the 1960s, the first aforementioned concern continued to be present at the highest levels of the USG. Mr. Allen Dulles, who was DCI from February 26, 1953 to November 29, 1961 (CNN 2021), is reported to have told former President John F. Kennedy, who served from January 20, 1961 to November 22, 1963, that the Central Intelligence Agency (CIA) had been sending decoy devices into Soviet airspace to test Soviet radar capabilities. In the course of these missions, CIA became aware that the Soviets could not distinguish aircraft and ballistic missiles from UAP. Dulles was concerned that the Soviets could mistake an unidentified object for a U.S. missile, and, on the basis of this error, could initiate a nuclear strike against the U.S. On November 11, 1963, the Soviets launched an unmanned spacecraft as part of their Kosmos 21 mission, but the spacecraft did not make it out of Earth's orbit. On November 12, 1963, the day after this unsuccessful rocket launch, Kennedy took two significant actions. He issued a National Security Action Memorandum instructing NASA to work with the Soviet space program, and he also issued a memo to CIA instructing it to share secret documents and other information with the Soviets on UAP. Ten days later, on November 22, 1963, Kennedy was assassinated, and his instructions to CIA to share UAP information with the Soviets were not carried out (UFOs 2019).

Former President Richard M. Nixon, who served from January 20, 1969 to August 9, 1974, shared the abovementioned concern that the Soviets might misidentify a UAP as a U.S. missile and accidentally start a war with the U.S. (UFOs 2019). In 1971, he and Soviet General Secretary Leonid I. Brezhnev signed the "Agreement on Measures to Reduce the Risk of Outbreak of Nuclear War Between the United States of America and the Union of Soviet Socialists Republics [USSR] – September 30, 1971." Article 3 of this agreement states:

> The Parties undertake to notify each other immediately in the event of detection by mission warning systems *of unidentified objects, or in the event of signs of interference with these systems or with related communications facilities*, if such occurrences could create a risk of outbreak of nuclear war between the two countries (Agreement 1971, italics added).

Reading the full text of this agreement underscores the significant concern held by both the U.S. and the USSR with respect to the potential for a nuclear exchange.

At least two former U.S. presidents—James ("Jimmy") E. Carter

Jr., who served from January 20, 177 to January 20, 1981, and Ronald W. Reagan, who served from January 20, 1981 to January 20, 1989—are reported to have had UAP sightings themselves. Carter has spoken openly about his sighting, which occurred in October 1969 when he was running for governor in Georgia. This sighting occurred outside the local Lions Club in Leary, Georgia, and was witnessed by approximately twenty people. Reagan's sighting is reported to have occurred in 1974 when he was flying in an airplane, and although Reagan himself never formally confirmed the sighting, the pilot of the aircraft did recount his version of the event. Speculation also exists that former President William ("Bill") J. Clinton, who served from January 20, 1993 to January 20, 2001, and/or his spouse, former Secretary of State Hillary D. Clinton, who served from January 21, 2009 to February 1, 2013—or a close family member or friend of the Clintons—also had a UAP sighting (UFOs 2019).

Quite recently, Ambassador R. James Woolsey, the DCI from February 5, 1993 to January 10, 1995, and Mr. John O. Brennan, the Director of the CIA from March 8, 2013 to January 20, 2017 (CNN 2021), both have indicated that an ETI may be responsible for some UAP operating on and around Earth. Brennan addressed the topic during two separate interviews (Brennan 2020; 2021). One of these interviews occurred on June 24, 2021, the day before ODNI released its *Preliminary Assessment*. During this interview, retired U.S. Army (USA) General Wesley K. Clark, who served as Supreme Allied Commander Europe of the North Atlantic Treaty Organization (NATO) from 1997 to 2000, asked Brennan about a number of topics, including UAP. Judging from Brennan's comments, it appears that his opinions with respect to UAP are based on photographic, video, and radar evidence to which he has been privy. Importantly, Brennan made statements during both interviews that indicate an openness on his part to the extraterrestrial hypothesis (ETH) as a way of explaining the origin of certain UAP.

Woolsey's comments on UAP indicate that he, too, has had access to the type of photographic, video, and radar evidence that

Brennan has seen, but that in addition to this, his opinions on UAP are informed further by the personal testimony of a close, trusted associate whose airplane experienced direct participation with an unidentified aerial object. I very much appreciate the position articulated by Ambassador Woolsey during his interview covering this topic when he stated that he hopes "that we can be friendly and able to deal with a wide range of behaviors in terms of dealing with our fellow human beings or with other creatures if they exist" (Woolsey 2021).

The views of both Woolsey and Brennan on UAP are very pertinent. In late 1993, while he was DCI, Woolsey ordered a review of all files on UFOs, and Brennan participated in this review. On September 30, 1993, Brennan wrote a memorandum entitled "Requested Information on UFOs," which he gave to Richard Warshaw, Executive Assistant to Woolsey at the time (G. Haines 1997). Brennan either was an analyst or an overseas case officer when he wrote this memo (Jackson 2009), presumably the former. Warshaw himself was a member of the External Referral Working Group (ERWG) established by the IC a few years later to confer and discuss former President Clinton's April 17, 1995 Executive Order 12958, Section 3.4 of which required that all permanent records twenty-five years or older be declassified by April 17, 2000 (FAS n.d.).

Nota bene, Hillenkoetter, Dulles, and Woolsey held the title "Director of Central Intelligence," whereas Brennan held the title "Director of the Central Intelligence Agency." This title change resulted from IC reform legislation passed in the U.S. in December 2004, which centralized IC leadership under the newly created Director of National Intelligence (DNI) while leaving a separate director in charge of CIA. The legislation provided the DNI with greater budgetary authority than the DCI had held previously, even though the DNI's authority to apportion funds did not automatically mean that the individual holding this position directed day-to-day operations. Prior to this, the NDAA for FY2003 had created the position of Under Secretary of Defense for Intelligence (USD(I)) to coordinate U.S. Department of Defense

(DoD) intelligence, to exercise authority over other important IC agencies including but not limited to Defense Intelligence Agency (DIA), National Geospatial-Intelligence Agency (NGA), National Reconnaissance Office (NRO), and National Security Agency (NSA), and to function as the point of contact between DoD and the DCI on resource and policy issues relating to intelligence. In keeping with these changes, the title for the head of the CIA changed from "Director of Central Intelligence" to "Director of the Central Intelligence Agency." Added specificity in the title reflected the establishment of the new DNI role, since the "Director of the Central Intelligence Agency" no longer exercised authority over the entire IC after the creation of this new DNI position (Best 2004, 2, 5, 8-9, 11, 18; see also Best 2010). The first person to hold the title "Director of the Central Intelligence Agency" was Mr. Porter Goss, who worked in this position from September 24, 2004 to May 5, 2006 (CNN 2021).

On May 18, 2021, former U.S. President Barack H. Obama II, who served from January 20, 2009 to January 20, 2017, stated with respect to UAP that "there is footage and records of objects in the skies that we don't know exactly what they are. We can't explain how they moved, their trajectory. They did not have an easily explainable pattern" (Obama 2021). On October 19, 2021, Mr. Clarence William ("Bill") Nelson II, a former U.S. senator and the current Administrator of the National Aeronautics and Space Administration (NASA), recently stated that U.S. Navy (USN) pilots have reported more than three hundred UAP sightings since 2004 (Nelson 2021).

Less than one month after Administrator Nelson made his comments, on November 11, 2021, the current DNI, Ms. Avril D. Haines, stated that UAP may have "come extraterrestrially" to Earth (A. Haines 2021). Less than two weeks after that, on November 23, 2021, Dr. Kathleen H. Hicks, Deputy Secretary of Defense (DepSecDef) at DoD, "direct[ed] the Under Secretary of Defense for Intelligence and Security (USD(I&S)) to establish the Airborne Object Identification and Management Synchronization Group (AOIMSG) to synchronize efforts across the Department [of Defense] and with other Federal departments and agencies to detect, identify and attribute" UAP in what the USG designates as Special Use Airspace (SUA) (Hicks 2021).

A November 23, 2021 memo from DepSecDef Hicks and a separate DoD press release of the same day both state that the AOIMSG will be established in DoD's Office of the USD(I&S) (Hicks 2021; U.S. DoD 2021). The NDAA for FY2020 redesignated USD(I), mentioned above, to Under Secretary of Defense for Intelligence and Security (USD(I&S)) (DeVine 2021). The Honorable Ronald S. Moultrie is the current Under Secretary of Defense for Intelligence & Security, and Mr. David M. Taylor is Performing the Duties of the Deputy Under Secretary of Defense of Intelligence & Security (OUSDI n.d.). In addition to tasking USD(I&S) to serve as the locus for AOIMSG, DepSecDef Hicks also tasked USD(I&S) "to lead an Airborne Object Identification and Management Executive Council (AOIMEXEC) to be comprised of DoD and Intelligence Community membership, and to offer a venue for U.S. government interagency representation" (U.S. DoD 2021).

In the aforementioned November 23, 2021 memorandum and also in the press release of the same day, DepSecDef Hicks draws attention to "Principal-level participation" of DNI Haines in AOIMEXEC. In the memorandum, Hicks writes:

> To provide oversight and direction to the AOIMSG, I establish the Airborne Object Identification and Management Executive Council (AOIMEXEC). The USD(I&S) will be the lead DoD official responsible for management of this process, will co-chair the AOIMEXEC along with the Director of Operations, Joint Staff, *and will invite Principal-level participation from the Office of the Director of National Intelligence* (Hicks 2021, italics added).

Similarly, the press release states that DepSecDef Hicks is taking the action to establish AOIMSG "*in close collaboration with the Director of National Intelligence*" [i.e., DNI Haines] (U.S. DoD 2021, italics added).

A little over a month after DoD's November 23, 2021 announcement that it was establishing AOIMSG and AOIMEXEC, the NDAA was signed into law on December 27, 2021 by U.S. President Joseph ("Joe) R. Biden Jr. (Singh 2021), who assumed office January 20, 2021. Section 1683(a) of the NDAA, ESTABLISHMENT OF OFFICE, calls for the establishment of an office to address UAP "within a component of the Office of the Secretary of Defense, or within a joint organization of the Department of Defense [DoD] and the Office of the Director of National Intelligence [ODNI]." Section 1683(h), ANNUAL REPORT, delineates the reporting requirements for this office. There, (1) states, "Not later than October 31, 2022, and annually thereafter until October 31, 2026, the Director, in consultation with the Secretary, shall submit to the appropriate congressional committees a report on unidentified aerial phenomena" (NDAA 2021, 578-79, 581).

Section (l)(1) of the NDAA defines "appropriate congressional committees" as follows: in the Senate, the Committee on Armed Services, the Committee on Appropriations, the Committee on Foreign Relations, and the Select Committee on Intelligence; and in the House of Representatives, the Committee on Armed Services (ditto Senate), the Committee on Appropriations (ditto Senate), the Committee on Foreign Affairs (corresponding to Senate), and the Permanent Select Committee on Intelligence (corresponding to Senate) (NDAA 2021, 583). Accordingly, from 2022 to 2026, eight congressional committees will receive annual reports on UAP.

According to Section 1683(b)(6) of the NDAA, AOIMSG is required to coordinate efforts to collect and analyze data with other departments and agencies of the USG. These include the Federal Aviation Administration (FAA), NASA, the Department of Homeland Security (DHS), the National Oceanic and Atmospheric Administration (NOAA), and the Department of Energy (DOE) (NDAA 2021, 579; for historical perspective, see also Huyghe 1979).

Because DoD in conjunction with ODNI already had called for the establishment of AOIMSG by the time Congress passed the NDAA and it was signed into law by President Biden, many people understandably were confused regarding whether or not the language of the NDAA indicated that a second office would be established within the USG to investigate UAP. This will not occur. Colonel Susan Gough, who is retired from the USA and who currently works as Senior Strategic Planner and Pentagon spokesperson, recently indicated that implementation guidance for AOIMSG will meet congressional intent regarding UAP as articulated in the NDAA, and that AOIMEXEC will assume an oversight role with respect to investigation of UAP events (Glassel 2021). Stating this more directly, by the time Congress passed the NDAA and President Biden signed it into law, it already was well established—at least in Washington, D.C.—that AOIMSG and AOIMEXEC would take the official lead within the USG on matters relating to UAP.

One point of commonality between the DoD/ODNI action and the NDAA is that both conclude and transition the work of the Unidentified Aerial Phenomena Task Force (UAPTF). Establishment of the UAPTF had been approved by Mr. David L. Norquist, then DepSecDef (the position Hicks now holds), on August 4, 2020. UAPTF had been established under the leadership of the Department of the Navy with cognizance of USD(I&S) (U.S. DoD 2020).

Now, the DoD's Memorandum of November 23, 2021 states that AOIMSG will supersede UAPTF: "Effective immediately, the AOIMEXEC, in coordination with the OSD [Office of the Secretary of Defense (OSD 2021)] and DoD component heads will manage the transition of the current UAP Task Force to the AOIMSG" (Hicks 2021, 1). Similarly, Section 1683(a) of the NDAA states that the new office mandated by Congress will "carry out the duties of the Unidentified Aerial Phenomena Task Force." In addition, Section 1683 (k) calls for the termination of the UAPTF "[n]ot later than the date on which the Secretary establishes the office under subsection (a)" (NDAA 2021, 578-79, 583). The foregoing means that the same office made responsible for UAPTF in 2020—namely USD (I&S)— now is responsible for AOIMSG and AOIMEXEC. Whereas UAPTF was a task force, however, AOIMSG is designated as a group.

As one may deduce from the discussion above, the topic of UAP is politically charged in the U.S. As with many other things, there is the official landscape—and then there is the actual landscape. Officially, DoD and ODNI are working together and are responsive to Congress on the topic of UAP. In actuality, three elements are important to this discussion: an alignment between U.S. Air Force (USAF) and CIA; an alignment between DoD and ODNI; and certain loosely-aligned members of Congress. This is somewhat confusing, however, since technically, USAF is part of DoD— but, on the topic of UAP, USAF is more closely aligned with CIA than with DoD's other elements. NASA negotiates this complex topography as well as it can, apparently taking appropriate direction primarily from CIA, and, also, from ODNI and DoD/ USAF, regarding what information it should or should not release with respect to UAP.

The USAF-CIA alliance with respect to UAP took shape quite early in the history of USG interest in the topic. Although it did arise out of an earlier organization active during WWII, namely the Office of Strategic Services (OSS), CIA was not officially founded in the U.S. until September 1947. Less than five years after CIA was formed, a special inquiry from the White House under President Harry S. Truman to CIA's Office of Scientific Intelligence (OSI) on August 14, 1952 led to the creation of a study group to investigate UAP. Shortly thereafter, the Assistant Director, Office of Scientific Intelligence, of CIA wrote a Memo dated October 2, 1952, "SUBJECT: Flying Saucers." It was addressed to the DCI, who at that time was Gen. Walter Bedell Smith, and it was conveyed to Smith by the Deputy Director (Intelligence). Under "2. FACTS AND DISCUSSION," the Memo states:

> OSI has investigated the work currently being performed on 'flying saucers' and found that the Air Technical Intelligence Center [ATIC], DI [Directorate of Intelligence], USAF, Wright-Patterson Air Force Base [WPAFB], is the only group devoting appreciable effort to this subject, that ATIC is concentrating on a

case-by-case explanation of each report, and that this effort is not adequate to correlate, evaluate, and resolve the situation on an overall basis (Aerial Phenomena 2021a).

During World War II, the USG relied heavily on technical intelligence. In July 1945, shortly before the war ended in September of that year, the Air Materiel Command T-2 Intelligence Department was founded at what then was called Wright Field (Wright Field and Patterson Field merged in 1948 to become WPAFB). One of T-2's responsibilities was to identify foreign aircraft and equipment. Not long thereafter, on May 21, 1951, ATIC was founded at WPAFB "as a field activity of the Assistant Chief of Staff for Intelligence." During the 1950s, ATIC analysts conducted computer analysis of aircraft in what was considered cutting-edge work for the time (NASIC n.d.). The fact that ATIC—a part of USAF—and CIA began to work together on the issue of UAP from around the time each was formed means that the USAF-CIA alignment on the topic of UAP goes back almost seventy years.

From the foregoing discussion, one may surmise that by 1952, at the latest, USAF already may have had one or more unidentified aerial objects or parts thereof in its possession at WPAFB. Today, the USG may continue to possess one or more unidentified aerial object or parts thereof, though not necessarily—or not necessarily exclusively—at WPAFB.

Section (h)(2)(K) of the NDAA describes what AOIMSG must report to the eight congressional committees delineated above annually until October 31, 2026. In particular, (K) states that the report shall include "[a]n update on any efforts underway on the ability to capture or exploit *discovered* unidentified aerial phenomena" (NDAA 2021, 581, italics added). Even though grammatically the sentence is ambiguous, the word "discovered" all but jumps off the page. One interpretation of the sentence is that if an unidentified ariel object is discovered, then the USG will attempt to "exploit" aspects of it, presumably its propulsion system, materiel composition, and/or other technological and/or materiel characteristics. Grammatically, however, the sentence also could mean that the U.S. already has in its possession at least one unidentified aerial object or parts thereof— i.e., that something already has been "discovered." *This is quite the humdinger.* This in fact may be true, even if this interpretation of the sentence was not intended (at least consciously) by whomever drafted the legislation.

Indeed, a USG historical document supports the interpretation that the U.S. already has in its possession 'discovered' UAP. On March 22, 1950, Guy Hottel, from SAC [Strategic Air Command] in Washington, D.C., wrote a memo to the Director of the U.S. Federal Bureau of Investigation (FBI), who at the time was Mr. J. Edgar Hoover. The memo states:

> An investigator for the Air Forces stated that three so-called flying saucers had been recovered in New Mexico. They were described as being circular in shape with raised centers, approximately 50 feet in diameter. ... [T]he saucers were found in New Mexico due to the fact that the Government has a very high-powered radar set-up in that area and it is believed the radar interferes with the controlling mechanism of the saucers (FBI 1950).

Hottel had been named acting head of the FBI's Washington Field Office in 1936. At the time he wrote the memo, he was special agent in charge (FBI 2013). As mentioned above, T-2 was an Air Material Command, so it would have made sense to send any recovered (i.e., "discovered") objects and/or parts to WPAFB given that ATIC analysts were spearheading the use of computers to analyze aircraft.

CHAPTER 2

State of Play

Given that DoD in conjunction with ODNI announced the establishment of AOIMSG and AOIMEXEC on November 23, 2021, prior to Congress passing the NDAA and it being signed into law by President Biden on December 27, 2021, one can conclude four things, at least. First, DoD, ODNI, USAF, and CIA will not compromise existing, classified policies and procedures relating to UAP. Second, DoD and ODNI do not think it is in U.S. strategic interests to curtail the longstanding USAF-CIA alignment with respect to UAP. Third, recent action by Congress articulated in the NDAA will not dislodge the longstanding USAF-CIA alignment on the topic of UAP and the overall support for this alignment within DoD and ODNI. Fourth, at the highest levels of the USG, USAF-CIA alignment on the topic of UAP is understood to be a critical element in maintaining the position of the U.S. in the global balance of power.

On the fourth point, above, I strongly support the USG's very reasonable decision to protect national security. Even more, I think this is the only reasonable position that can be taken at the current time. The four elements within the USG working together here—DoD, ODNI, USAF (again, recognizing that technically USAF is part of DoD), and CIA—all are demonstrating considerable integrity in exercising appreciable caution regarding what information can or cannot be released with respect to UAP. My view is that all of these entities are determined—commendably —to avoid a dystopian outcome in this era of what many are referring to as great power competition.

As I read the subtexts of many conversations occurring in and around Washington, D.C., the legitimate concern with respect to any potentially existing classified information on UAP propulsion and/or materiel composition, in particular, is that a peer adversary of the U.S., e.g., the People's Republic of China, or PRC (henceforth China), and/or the Russian Federation (henceforth Russia), and/or some other country and/or group, were to effect a strategic reversal by gaining access to such information and then using this information to attempt to unseat the U.S. in the global balance of power. This is all the more important now when Russia and China have declared that they are operating as a bloc (President of Russia 2022). My view is that this stance on the part of the U.S. is *even more imperative now* following an oblique comment made by Russian President Vladimir Putin in the midst of its extreme aggressivity toward Ukraine with respect to Russia's own weapons development initiatives:

> We will continue to develop advanced weapon systems, including hypersonic and those *based on new physical principles*, and expand the use of advanced digital technologies and elements of artificial intelligence.

> Such complexes are truly the weapons of the future, which significantly increase the combat potential of our armed forces (Lock 2022, italics added).

Depending upon what is meant by the italicized portion of this quotation, one develops even greater understanding for why individuals in the U.S. are taking the strongest possible steps to safeguard any classified information that may exist on UAP propulsion and/or materiel composition.

The values on which our democratic system in the U.S. is founded are admirable. Individuals within the USG rightfully are concerned that if a foreign adversary such as China or Russia—and/or an extremist group— were to gain access to a paradigm changing technology such as the one undergirding UAP propulsion, that such an actor or actors could replace the U.S. political system and its democratic principles with a political system based on authoritarian principles, autocratic leadership, and repressive, social control. This is an imminently reasonable position concern to have and in the context of Russia's combativeness and expansionism and the calculating manner in which China is eyeing Taiwan.

In this era of multiple nuclear risks and brinksmanship diplomacy, it is vital that we get our national security priorities straight. Pragmatically, I support a traditional national security posture that safeguards whatever classified information on UAP propulsion and/or materiel composition may exist while humankind develops the ethical sensibility and spiritual mooring necessary to handle this information safely.

Even in a democracy such as ours in the U.S., which prides itself on being an open society, pragmatic protections are built into our legal system in order to safeguard particularly sensitive information. It is the appropriate, legal purview of the Secretary of Defense [SecDef] to decide not to release certain information widely, even to the full Congress, when such information is deemed essential to national security. This in no way constitutes illegal concealment of information.

Assuming for a moment that the USG holds classified information on UAP, and, moreover, and UAP propulsion and/or materiel composition, in what often is referred to as a 'waived SAP,' then details regarding such a SAP legally are waived from broad disclosure to Congress. This is made clear in the statute governing SAPs, namely 10 U.S.C. [U.S. Code] § 119, Special access programs: congressional oversight. According to Section (e)(1) of the law:

> The Secretary of Defense may waive any requirement ... that certain information be included in a report ... if the Secretary determines that inclusion of that information in the report would adversely affect the national security. Any such waiver shall be made on a case-by-case basis (Special access programs 2011, 2).

However, even when the SecDef legally decides to leave certain information out of reports to the full Congress, waived SAPs still are subject to Congressional oversight. Section (e)(2) of 10 U.S.C. § 119 requires the SecDef to provide information on each waived

SAP "jointly to the chairman and ranking minority member of each of the defense committees." Section (g) defines the term "defense committees." There, (g)(1) states that in the Senate, the relevant committees and subcommittee are "the Committee on Armed Services and the Committee on Appropriations, and the Defense Subcommittee of the Committee on Appropriations." Section (g)(2) states that in the House of Representatives, the relevant committees and subcommittee are "the Committee on Armed Services and the Committee on Appropriations (ditto Senate), and the Subcommittee on Defense of the Committee on Appropriations" (corresponding to Senate) (Special access programs 2011, 2). As the law makes clear, the aforementioned, designated members of Congress possess oversight of waived SAPs even when the SecDef decides that national security concerns are so significant that only this select group in Congress should be briefed. Nothing extrajudicial is occurring here, since the very purpose of waived SAPs is to provide a mechanism whereby particularly sensitive information can be safeguarded for the purpose of robust national security.

Assuming again for the sake of discussion that information on UAP propulsion and/or materiel composition exists in a waived SAP, I am in favor of keeping it there at least at this juncture in human history—not only to protect the position of the U.S. in the global balance of power, but, also, to preserve the existence of humankind. What do I mean by this? UAP have the ability to access incredible amounts of energy, which some people measure theoretically according to the Kardeshev Scale (Creighton 2014). While millions of human beings—maybe even tens of millions—are ethically and spiritually advanced enough that they would use virtually unlimited energy for the good of the entire species and the planet, this simply is not true of all human beings. Many people are profoundly unkind, to the point of being manipulative, ruthless, and violent. It is quite possible that such individuals would try to exploit the vast energy associated with UAP to benefit themselves and/or their close associates, and, furthermore, to exert repressive social control over others. Even worse, if such individuals were in the throes of a deep-seated delusion—e.g., a genre of millenarianism—then they could attempt to use the vast energy source associated with UAP to create an apocalyptic scenario in the misguided attempt to usher in some imagined next stage. That would be a recipe for the obliteration of humankind and much else with it. I therefore support all elements within the USG that are taking a prudent approach to keeping any information on UAP propulsion and/or materiel composition appropriately classified at the highest level until such segments of human society come to their senses. To be clear, mine is not an argument for secrecy—it is an argument for survival.

I also think it is helpful to differentiate openness from transparency. Openness is a positive, democratic value that citizens must uphold for the sake of preserving and expanding civil liberties in a civil society. Civil society depends upon open access to education and a civilized climate in which political and social issues can be discussed openly without fear of stigmatization, bullying, and/or retribution. I fully support openness with respect *to the existence of* UAP and ETI, and I express my own view on this topic very openly. I do this because I think it is important for humankind to assimilate the existence of an advanced ETI present around us as expeditiously as possible (Andresen 2021), for psychological, ethical, and spiritual reasons (Andresen 2023; in preparation). I actually think the USG could take another step forward in this regard by stating openly that on the basis of information it has in its possession, extraterrestrial origin is the most likely—if not only—explanation for many UAP.

While I support openness with respect to the existence of UAP and of ETI, I nevertheless do not support transparency with respect to classified information on UAP propulsion and/or materiel composition. Putting this quite bluntly, I think such transparency would be insane given the current geopolitical landscape. Here, I am making a pragmatic calculation of risks in conjunction with how many people do or do not have access to this information. Assuming again for the sake of discussion that a waived SAP with information UAP propulsion and/or materiel exists, then I do not think that the persons involved with such a program necessarily are the only individuals on Earth who would handle such information responsibly. But this is a far smaller problem than potential misuse of this information — again, because of the incredible amounts of energy to which UAP have access. We only have to look for comparison at how appallingly humankind has handled nuclear energy while realizing simultaneously that the amount of energy to which UAP have access makes nuclear energy look elementary. Why do you think ETI has been so patient in releasing information to humankind? Could it be that ETI does not wish to have its own knowledge exploited by nefarious and deeply deluded human actors who might attempt to use it to annihilate the entire human species? In light of these considerations, my pragmatic position is that it is advisable for the USG to retain whatever information it may possess on UAP propulsion and/or materiel composition at the highest possible level of classification, with the fewest number of persons whomsoever having access to it. This position can be revisited at a future time in humanity's history, when it is safe to do so. Right now it clearly is not.

Time matters. What we need is a careful, judicious, and step-by-step approach to releasing information on UAP propulsion, etc., that unfolds at the pace at which humankind becomes ethically and spiritually capable of handling this information (Andresen 2023). Ours must be a cogent, long-term view. As human beings become less vicious and less violent, as we become kinder and more gracious to one another, to other species, and to the world around us—and as we move past the mindset in which people and situations are used transactionally and exploitatively to maximize one's own personal advantage—then human beings will be able to share whatever information they already may possess on UAP propulsion and/or materiel composition and, also, ETI will be able to share more information safely with humankind with respect to its own technologies. This will be a paradigm changer for our species. We simply must ensure that this knowledge and the intense amount of energy associated with it is not used destructively, but, instead, that it is used to benefit humankind, other species, and the environment as a whole.

CHAPTER 3

A Bohm's-Eye View

Here, I propose a deductive approach to understanding ETI and its UAP. A deductive methodology proceeds differently than an inductive one. Inductive reasoning usually proceeds on the basis of unintentional yet flawed incorrect and reductive assumptions that form misinformation at a central position or elsewhere in a system. Such misinformation arises on the basis of historical happenstance and then is transmitted throughout the system. With respect to trying to improve the map of academic disciplines, for example, an inductive approach would add, redraw, and/or erase lines between disciplines. It would miss entirely the fact that the stability one sees along the main contours of the academic map—rather being an indication of the map's correctness—masks significant intellectual rigidity and social stagnation underneath.

Today, misinformation is held rigidly in human society today by individuals in many walks of life, including those in the academy and in government. To remediate the negative impact of misinformation, one can reason deductively from the broadest possible framework discernible at any point in time. As a useful starting point for such deductive reasoning, I use the ontological interpretation of quantum mechanics outlined by theoretical physicist and philosopher David Bohm.

As Bohm observes, "misinformation at a central position" can result in serious, negative consequences. Bohm makes an analogy to the DNA molecule: "A tiny bit of misinformation in the DNA can cause everything to go wrong. If there is misinformation at a crucial part of the information that is determining a process, the consequences can be very serious" (Bohm 1989, 23).

Bohm differentiates information from misinformation. He

uses a virus as an example of misinformation, stating, "A virus in the DNA molecule could be called a bit of misinformation, in that it enters the genetic structure and causes the cells to produce more viruses instead of more cells." He continues, "Society, too, is full of this sort of misinformation. We have various ways of dealing with biological misinformation. The best way is by the immune system which recognizes it and gets rid of it, but we have no such system in society. Misinformation accumulates and society gradually decays." Bohm comments that as a society becomes older, it has more chance "to accumulate all sorts of misinformation." If this occurs, the society begins to fall apart. He writes, "The society is blocked because misinformation is held *rigidly*" (Bohm 1989, 22).

Because an inductive approach would involve misinformation, I recommend that we take a deductive approach to ETI and UAP based on Bohm's ontological interpretation of quantum mechanics. Such an approach is particularly helpful in illuminating how the ETI responsible for extraterrestrial UAP operating on and around Earth perceives and processes information.

Witness reports and other data regarding many UAP events indicate that extraterrestrial UAP can maneuver in remarkable ways. This suggests to me that the ETI responsible for these UAP perceives, understands, and can operationalize what Bohm refers to as the inseparability of consciousness and matter and, also, the wholeness of reality. With respect to the inseparability of consciousness and matter, unlike human beings, who have a penchant for dichotomizing, the ETI operating UAP on and around Earth now does not seem to process information according to any permutation of dualistic cognition at all. Perceiving, understanding, and operationalizing the inseparability of consciousness and matter enables this ETI to avoid "misinformation at a central position" (Bohm 1989, 23).

My view is that at least one of ETI's objectives in interacting with humankind is to help human beings progress beyond its many misperceptions of and misinterpretations relating to the nature of reality. Maladaptively, human beings currently base much of their thinking and actions on the misinformation that consciousness and matter are distinct. Now, imagine an ETI that is not impeded by misinformation in the form of consciousness/ matter dualism. Maybe members of this extraterrestrial species never thought dichotomously at all, since they may not be embodied in the same way that human beings are, members of the species may be genetically and/or otherwise engineered to avoid cognitive reductionism, and/or something else we have not thought of yet. Or, members of this extraterrestrial species may have developed a general pattern of dichotomous thinking as an epiphenomenon of proprioception, navigation, etc., then later moved beyond this maladaptive habit to a non-dichotomous cognitive style. Bohm eloquently discusses the inseparability of consciousness and matter in the context of his ontological interpretation of quantum mechanics, contextualizing this concept within of his broader explanation of holomovement and implicate and explicate levels of order (Bohm 1952; 1980; 1987; 1990; Bohm and Hiley 1987; 1995). Holomovement is "the ground of what is manifest," and its "basic movement is folding and unfolding" (Bohm 1986, 26). Here, "what is manifest" refers to explicate order. Bohm summarizes his view as follows: "The essential feature of this idea was that the whole universe is in some way enfolded in everything and that each thing is enfolded in the whole." Further, "this enfoldment relationship is not merely passive or superficial. Rather, it is active and essential to what each thing is. It follows that each thing is internally related to the whole, and therefore, to everything else" (Bohm 1990, 273).

While human beings are socialized into, and, therefore, accustomed to thinking of reality as the display of so-called explicate order, this is only one aspect of reality. Bohm writes:

> The external relationships are then displayed in the unfolded or explicate order in which each thing is seen, as has already been indicated, as relatively separate and extended, and related only externally to other things. The explicate order, which dominates ordinary experience as well as classical (Newtonian) physics, thus appears to stand by itself (Bohm 1990, 273).

Bohm insightfully observes that the ground of explicate order is "the primary reality of the implicate order." Furthermore, the relationship of implicate and explicate orders accounts for "the meaning of the properties of matter, as implied by the quantum theory" (Bohm 1990, 273).

Bohm's ontological interpretation of quantum mechanics includes the concept of active information. Bohm uses the analogy of a ship on automatic pilot that is guided by radar waves to explain active information:

> The ship is not pushed and pulled mechanically by these waves. Rather, the *form* of the waves is picked up, and with the aid of the whole system, this gives a corresponding shape and form to the movement of the ship under its own power. Similarly, the form of radio waves as broadcast from a station can carry the form of music or speech. The energy of the sound that we hear comes from the relatively unformed energy in the power plug, but its form comes from the activity of the form of the radio wave; a similar process occurs with a computer which is guiding machinery. The 'in-formation' is in the programme, but its activity gives shape and form to the movement of the machinery (Bohm 1990, 279).

The process described above is analogous to a living cell, observes Bohm, since "the form of the DNA molecule acts to give shape and form to the synthesis of proteins (by being transferred to molecules of RNA)" (Bohm 1990, 279).

Bohm extends the notion of active information to matter at the quantum level. He writes:

> The information in the quantum level is potentially active everywhere, but actually active only where the particle is (as, for example, the radio wave is active where the receiver is). Such a notion suggests, however, that the electron may be much more complex than we thought (having a structure of a complexity that is perhaps comparable, for example, to that of a simple guidance mechanism such as an automatic pilot) (Bohm 1990, 279).

Here, Bohm's view differs from traditional physics views that hold that as one analyzes matter into smaller and smaller parts, the behavior of these parts becomes simpler and simpler. Bohm points out that such assumptions sometimes are mistaken. For example, Bohm notes that large crowds of people often display a much simpler behavior than that of the individual human beings who comprise them (Bohm 1990, 279).

One implications of the notion of active information is quite profound, namely that matter possesses a mind-like quality. Bohm states:

> [T]he whole notion of active information suggests a rudimentary mind-like behaviour of matter, for an essential quality of mind is just the activity of form, rather than of substance. Thus, for example, when we read a printed page, we do not assimilate the substance of the paper, but only the forms of the letters, and it is these forms which give rise to an information content in the reader which is manifested actively in his or her subsequent activities (Bohm 1990, 281).

As one example of the notion of active information, Bohm comments that "ballet-like" behavior in superconductivity in the quantum theory "is clearly more like that of an organism than like that of mechanism" (Bohm 1990, 281).

In human thinking, consciousness/matter dualism manifests as two extremes—idealism, which results from reducing to the mental pole; and materialism, which results from reducing to the physical pole (Bohm 1990, 272). Such reductionism is evident in the activities of all institutions in human society, be they academic, governmental, corporate, religious, etc. Bohm often uses the terms mind and matter rather than consciousness and matter. I tend to uses the terms consciousness and matter because I usually reserve the term mind for a special case of consciousness/matter dualism, namely mind/body dualism. In an analogous manner, mind/body dualism and consciousness/matter dualism both are incorrect.

According to Bohm, thoughts have a physical aspect and physical things have a mental aspect. Taking thought as an example of how the notion of active information helps explain the inseparability of mind and matter, Bohm states that a major aspect of the significance of thought is "the activity to which a given structure of information may give rise." Even though human beings often perceive information in thought to be mental, thought simultaneously possesses physical aspects that are neurophysiological, chemical, and that relate to physical activity. For example, if an individual who encounters shadows a dark night possesses the idea that enemies lurk in the neighborhood, this person may experience mental processes and, also, processes that are involuntary and essentially unconscious, such as hormones, heartbeat, neurochemicals, physical tensions, movements, etc. Therefore, no strict division exists between the mental and physical aspects of reality, and one understands that "with mind, information is thus seen to be active in all these ways, physically, chemically, electrically, etc." In this sense, Bohm notes that active information "is simultaneously physical and mental in nature" and that it functions as a bridge between these inseparable aspects of reality (Bohm 1990, 281-82).

According to Bohm, an infinite number of levels of subtlety exist both mentally and physically. He writes:

> We have thus been led to an extension of the notion of implicate order, in which we have a series of inter-related levels in which the more subtle—i.e. 'the more finely woven' levels including thought, feeling and physical reactions—both unfold and enfold those that are less subtle (i.e. 'more coarsely woven'). In this series, the mental side corresponds, of course, to what is more subtle and the physical side to what is less subtle. And each mental side in turn becomes a physical side as we move in the direction of greater subtlety (Bohm 1990, 282-83).

With respect to the aforementioned series of levels, the mental side is *potentially active information*. Whereas human thought contains a range of information content of different kinds that can be perceived by a higher level of mental activity as if it were physical, a more subtle level of information may emerge from this. The meaning of this more subtle level of information is an activity that can organize the original set of information into a greater whole. Similarly, a more subtle level of mental activity can survey this more subtle information, and this process in principle can continue indefinitely. From the material, i.e., physical, side, an actual activity organizes the less subtle levels, which serve as the so-called material on which such an operation occurs. A similar relationship occurs "at indefinitely great levels of subtlety," and at each level, "information is the link or bridge between the two sides" (Bohm 1990, 282).

CHAPTER 4

Proposing A Continua-Based Phenomenology for UAP

In keeping with the notion of the inseparability of consciousness and matter, I do not subscribe to an ontological separation between objective/quantitative and subjective/qualitative forms of information. Notwithstanding this caveat, it is possible, albeit only provisionally, to organize the vast amount of information available on UAP into objective/quantitative data (radar reports, photographs, videos, etc., collected in the field, and, also, measurements, graphs, etc., created by researchers) and subjective/qualitative data (testimony of firsthand witnesses to UAP sightings, reports from individuals reporting direct encounters with ETI, etc.). That being said, one should not reify one's ideas regarding what constitutes so-called objective information and what constitutes so-called subjective information but, instead, one should remain open to a deeper, ontological understanding of the inseparability of consciousness and matter.

Tremendous variety in UAP reports made by numerous witnesses around the world demonstrates that UAP can manifest in many different ways. The question therefore arises as to whether or not we should try to create a working typology of UAP. Indeed, it is a natural aspect of the human cognitive process to try to organize things. We do this at almost every level, from organizing chemical elements in the periodic table to organizing denominations within religions. Similarly, we organize topics of study in the academy into disciplines, subdisciplines, etc. We even try to organize people into groups based on all sorts of contrived divisions.

In fact, the human compulsion to categorize things is

apparent in the many attempts to categorize UAP that have been made over the years. For example, shortly after the Washington D.C. UFO flap of July 1952 (Aerial Phenomena 2021d; UFOs 2019) and in that same year, CIA created a typology that organized UAP according to size and speed. This typology described UAP as small, medium-sized, and very large, and it categorized their speeds as "hovering," "moderate," and "stupendous" (Aerial Phenomena 2021a). In contrast to CIA's quantitative typology, UAP investigator and writer John A. Keel's two-part typology for UAP is more qualitative, categorizing UAP as "hard" and "soft" (Keel 1969). Because it moves beyond a simple quantitative approach based on estimates of sizes and speeds, Keel's typology helps one think about UAP and related phenomena more impressionistically.

But, what if it is misguided to try to typologize UAP at all? The proclivity of the human mind to organize things into categories may be obscuring the real nature of UAP. This would be the case if UAP do not exist as rigidly defined 'types,' but, instead, if they exist as interpenetrative continua that transcend the perceptual and conceptual dichotomy between 'object' and 'non-object.' Indeed, I think that UAP do exist as interpenetrative continua, and that everything in fact exists in this manner— which both Bohm and ETI are trying to elucidate.

My view is that the complex nature of UAP and related phenomena possesses meaning, and that ETI intentionally uses magnificent displays of maneuverability to teach humankind something profound regarding the nature of reality. The complexity of UAP displays stimulates the human brain to think, and it also catalyzes people to collaborate together to try understand the nature of UAP. Considering the situation from this vantage point, one quickly realizes that ETI has extremely advanced pedagogical skills. Adding some mystery to UAP events often makes them even more compelling, such that they become Brobdingnagian pedagogical tools. Here, ETI really excels.

Bohm recognizes the lack of inherent existence of any apparent 'thing' in his concept of holomovement, pointing out that implicate and explicate are only poles along a continuum, and, similarly, physical and mental are only poles along a continuum. Neither pole exists in a reified manner, since the concept of a pole is abstracted only for the purpose of explanation. Something analogous is occurring with UAP, which means that any rigid categorization of UAP self-deconstructs. It therefore makes sense to think about the nature of UAP—indeed, the nature of everything—as existing as interpenetrative continua.

A continua-based phenomenology of UAP has practical value in describing the appearances of UAP from the vantage point of human observers. I label the two poles of the continuum I am proposing here as *defined* and *amorphous*. These two terms are apt linguistically because they connote intentionality, and I think ETI intentionally is signaling something important about the nature of reality to human beings when it manifests intriguingly in a manner that defies categorization. Recognizing that defined and amorphous are terms describing opposite poles along a continuum and are not rigid categories in and of themselves, one nevertheless can describe how UAP appear and act at each of these poles. This is not unlike describing how something feels or is experienced physically versus mentally, even when one has assimilated Bohm's insight that thoughts have a physical aspect and physical things have a mental aspect.

At the defined pole of the continuum, UAP appear and behave craft-like and mechanical. Here, UAP also appear as if they have, or at least could have, beings inside them. Toward the defined pole of the continuum, UAP include discs, often described as silver, white, metallic, and/or bright red; saucers; triangles; cylinders; spheres; cubes; cubes within spheres; circles within circles; cones; and black lenticular UAP with blinking lights on the surface. Some of these shapes, along with others, are depicted in a "UFO Identification Chart" from 1967 (ArtisMortisGallery 2021). Also included here are objects that witnesses liken to the shapes of known objects such as eggs, acorns, *unshelled* walnuts, bowler hats, boomerangs, diamonds, radio antennae, Tic Tacs, six-pronged stars, doughnuts, Dropa stones, etc. This list of descriptions is not exhaustive, and it merely delineates some commonly described UAP shapes. One immediately recognizes that a saucer and a bowler hat have roughly the same shape, and that a doughnut and a Dropa stone have roughly the same shape —though in the second example, the first witness presumably was hungry while the second witness sounds like an anthropologist. It therefore is important to realize that two witnesses may employ different words or phrases to describe the same underlying object or phenomenon.

At the amorphous pole of the continuum, UAP do not appear to have beings inside, at least not the kind of sentient beings with whom humans are familiar, i.e., embodied ones. Here, one finds UAP appearing as round or oval balls of light; orbs; round or circular glowing orbs that are reddish orange or glowing red; plasma-like orbs; triangular, plasma-like objects; orange lights that change in number; bright red, pulsating objects in an octagonal formation; white, luminescent objects; fireballs, sometimes described as fiery balls; gaseous spheres; etc. UAP in the form of question marks even have been observed (Keel 1969, 27), as if to say "Dear Humans, What *are* you thinking?"

Intriguingly, sometimes during what appears to be a single UAP display, witnesses observe UAP with both defined and amorphous elements occurring together and/or in close spatial and/or temporal proximity to one another. For example, some plasma-like UAP appear to manifest somewhere in the middle of the defined and amorphous poles of the continuum, though probably closer to the amorphous pole. While having an indistinct, plasma- like appearance, however, these UAP appear somewhat like objects that at least theoretically could carry occupants. In what is a real stonker, too, UAP can appear to change shape (Aerial Phenomena 2021c), a capability described spectacularly as "transmogrification" (Keel 1969).

Analogous to the abovementioned non-dichotomous aspect of UAP and in keeping with Bohm's observations regarding the inseparability of consciousness and matter, some UAP appear both mental and physical at the same time. In other words,

UAP simultaneously can display characteristics that are both defined and amorphous and/or physical and mental within the parameters of what observers interpret as a relatively discrete event. This means that UAP have the ability to ebb and flow effortlessly between poles along the defined/amorphous continuum, thereby clearly demonstrating that the ETI responsible for these UAP is not stumbling over an incorrect dichotomy between consciousness and matter.

The fact that UAP can move between defined and amorphous poles of the continuum I propose here suggests that UAP encompass all aspects of the electromagnetic spectrum—which includes radio, microwave, infrared, visible, ultraviolet, X-ray, and gamma ray elements. When witnesses comment on the colors of UAP, for example, this observation relates to human perception within the visible component of the electromagnetic spectrum.

Very engagingly, toward the amorphous pole of the continuum, UAP often appear and/or are experienced by witnesses as if they are sentient—or at least as if a sentience of some form is observing from within and/or beyond the actual UAP themselves. This means that the farther along the overarching defined/amorphous continuum one moves towards the amorphous pole, the more redundant it would be for beings to be 'inside' UAP, since the UAP themselves already display aspects and/or degrees of sentience. In addition, the sentience displayed at the amorphous pole of the continuum is very advanced. Furthermore, in a meta manner, sentience itself also appears to manifest as a continuum. I call this the *sentience spectrum*. At the upper reaches of subtlety, sentience often manifests a pearlesque and ethereal quality that is impossible fully to evoke linguistically.

Manifesting as interpenetrative continua instead of as rigid types, UAP sometimes appear elusive to human observers. I do not think we should read anything into this, however. What we should do is consider that typologizing UAP in any way may not be as applicable, useful, and/or meaningful as we think, since it simply may be the nature of the phenomenon not to exist in a rigid manner. In addition, one of many interesting aspects of

UAP is that they often are observed flying in formation (Toftenes and Toftenes 2010; Aerial Phenomena 2021c; Proof 2021). Indeed, I often think that UAP are *flying information*—and that the 'in formation' aspect of their movement encodes a special type of information that transcends the simple observation that more than one unidentified object *often appears to appear* at one time. Based on the adroitness of ETI's UAP to maneuver, my view is that the ETI operating on and around Earth now can operationalize multiple organizational levels of increasing subtlety (e.g., implicate order, super- implicate order, super-super-implicate order, etc.) in the holomovement, potentially even extending to infinity. For example, the ability of UAP to appear and disappear at will demonstrates that ETI possesses a profound understanding of and ability to operationalize reality by moving seamlessly within the holomovement to manifest, unmanifest, manifest, etc. Accordingly, reflecting upon reality only from an explicate order perspective is
insufficient, since one also must include implicate, etc., order awareness.

CHAPTER 5

ETI's-Eye View

I am certain that the ETI responsible for extraterrestrial UAP operating on and around Earth is not a threat to humankind. Despite ETI's longstanding connection to humankind, and despite possessing the ability to utilize incredible amounts of energy, the ETI around us here on Earth has not prevented the human population from growing to almost eight billion people. Natural resources are abundant in space, and there are many exoplanets to choose from if a species desires a terrestrial abode. For an advanced ETI, there is no reason whatsoever to designate Earth as a 'must have' planet and then go about exterminating its citizens. It makes no sense, and it is not occurring.

The very fact that an ETI is here—or that any ETI may arrive in the future—should provide human beings with significant reassurance that we are not dealing with a hostile ETI now, and that we will not be dealing with a hostile ETI in the future. Why? Because simply to travel to Earth, any ETI will have mastered use of an almost incomprehensible amount of energy while at the same time possessing a holistic enough worldview that it has not wiped itself out. No species can master virtually unlimited energy safely and survive without realizing how interconnected everything is. In fact, a genuinely holistic perspective on reality *inherently is nonthreatening*, since rather than certain elements of a system being valued above others, existence itself is perceived and valued as a whole. This relates directly to the notion of intelligence itself.

What are the qualities of intelligence? As Bohm insightfully recognizes, "the essential feature on which the possibility of intelligence is based" is the possibility of moving "beyond any specifiable level of subtlety" (Bohm 1990, 282). This helps us understand how advanced the ETI responsible for the extraterrestrial UAP operating on and around Earth actually is. This ETI has moved far beyond the level of subtlety (or, more aptly, coarseness) at which human consciousness routinely operates.

In contrast to ETI's demonstrated sophistication, humankind often is deluded about the degree of its own intelligence. To ETIs, however, human beings must appear profoundly immature in their impulses and outbursts, particularly those that hurt and kill other human beings, eradicate other species, and damage the environment. I possess the strong conviction that the ETI around us now cares very deeply about social justice and sustainability and is intent on helping humankind move past recurring, negative patterns and networks of domination and control that have been part of human society for centuries. In contrast to humankind's own negative behavior patterns, ETI is not controlling human beings—certain human beings are controlling other human beings using many different strategies. ETI clearly is attempting to help humankind move past this morass.

One thing that ETI seems to have determined correctly about human beings is that we are extremely bad at managing high amounts of energy. A case in point is the nuclear situation on Earth. Indeed, one should not have to be a supersmart extraterrestrial to realize that nuclear weapons are a bad idea, and that nuclear technologies of any type are too dangerous to be prudent. Human beings should be able to figure this out for themselves. Sadly, however, as a species, we have not. The nuclear crisis is particularly acute now because of increasingly tense global competition, hostilities between nation-states and nonstate actors, and engagement by multiple parties in dangerous military practices.

ETI has good reason to be concerned about humanity, which is at the brink of annihilating itself as a species, potentially taking other species and vast swaths of the natural world with it. I call this the *precariousness quotient* (PQ), which is the antithesis and nemesis of the intelligence quotient (IQ) (Andresen 2023).

The baseline aspect of what I call the PQ with respect to nuclear technologies is what Krepon (2021) refers to as the geometry of competition involving nuclear-armed rivals. This geometry is far more complex than Krepon even states. Almost assuredly, ten countries already possess nuclear weapons. Eight of these countries are overtly nuclear—Russia, U.S., China, France, U.K., Pakistan, India, and North Korea. The first seven countries are listed in the order of greatest to least total number of deployed plus stored warheads. North Korea's position on the list refers to the number of warheads it is estimated the country could build with the amount of fissile material the country is estimated to have produced (Kristensen et al. 2021, 2). In addition, two other countries almost certainly possess nuclear weapons, Israel (Kristensen and Korda 2022; Trevithick 2019; Gross 2021; and Cordesman 2005, 49) and Iran (Woolsey et al. 2016; 2021), although neither country overtly claims to do so. With the assistance of China, Saudi Arabia also may be manufacturing its own ballistic missiles (Cohen 2021). A nuclear nation also theoretically could supply a nuclear weapon to a proxy nation and/or to an ally, further complexifying the geometry involving nuclear-armed rivals.

En route to less conflictual world that renders the notion of national security unnecessary, we must increase the amount of money we spend on nuclear nonproliferation and reduce what is spent on weapons development. In dollar terms, the NDAA allocates $777.7 billion in spending for so-called national defense, of which $740.0 billion is allocated to DoD and $27.8 billion is allocated to so-called national security programs within DoE (U.S. SASC 2021). More precisely, the NDAA authorizes DoE to spend $27,813,592 in discretionary funding, with $15,981,328 of this amount authorized for "Weapons Activities." In comparison, a meager $1,957,000 is authorized for "Defense Nuclear Nonproliferation" (NDAA 2021, 795- 801). This means that only 0.25% of the NDAA is allocated for nuclear nonproliferation in Fiscal Year 2022. In other words, the U.S. will spend only a quarter of 1% of the NDAA's overall budget on the pressing issue of nuclear nonproliferation. Given the current geopolitical climate, this is insane.

The lethality of nuclear weapons also is extremely problematic, and, like the baseline geometry of competition involving nuclear-armed rivals, also factors into the PQ with respect to nuclear technologies. Some of the nuclear weapons today are at least 3,000 times more powerful as the bomb dropped on Hiroshima, which killed between 90,000 and 166,000 people (Bennett 2020). In addition, nuclear weapons are becoming more destructive and riskier because of recent engineering modifications (Smith 2021). Furthermore, agreement on a Fissile Material Cut-off Treaty (FMCT) has been impeded because the consensus model of the United Nations (UN) Conference on Disarmament (CD) has resulted in stagnation (Kimball 2018; NTI 2020).

Global stockpiles of highly-enriched uranium (HEU) and of separated plutonium have reached horrific levels (IPFM 2021; IPFM 2015, 2). India produces plutonium for weapons and HEU for naval propulsion, Pakistan produces plutonium and HEU for weapons, it is thought that Israel is producing plutonium, and North Korea can produce both weapons-grade plutonium and HEU (IPFM 2021). In addition, Iran began stockpiling enriched uranium in 2019 (IAEA 2021; Iran International 2021).

It is not merely that human beings 'possess' nuclear weapons, either. We live in a world in which thousands of nuclear weapons *are armed and pointed* at every major population center on Earth, and at all of the critical infrastructure required to keep human society functioning. This includes power grids, freshwater pipelines, wastewater systems, highways, bridges, tunnels, railways, airports, utilities, food distribution hubs, grocery stores, financial centers, IT networks, communications systems, etc., etc., etc. All of it is rigged to explode in the event of widespread nuclear holocaust.

Planetary dynamics and armed conflict make the PQ associated with nuclear technologies even more serious. Tectonic shifts, earthquakes volcanic eruptions, asteroid and/or comet impacts, and other events all have the potential to trigger a nuclear accident. Approximately 20% of the nuclear reactors in the world operate in earthquake danger zones (Randall 2014). Climate change, including coastal flooding, further intensifies risks associated with nuclear installations (Polansky 2018; USA War College 2019). Potential flooding from upstream dam failures also poses a serious risk to nuclear power plants (Perkins et al. 2011). These installations also can be seized during armed conflicts, resulting in damage, accidents, and potential mass destruction events.

My view is that the firmer presence in recent years of UAP around U.S. military targets relates directly to the risk that nuclear weapons and other nuclear technologies pose to humankind. In typical ETI meta fashion, ETI is shining a light both literally and metaphorically on the precariousness of the nuclear situation on Earth by having its UAP appear at nuclear sites and—literally—shine a light on them. Human beings should be intelligent enough to realize just how much wisdom and, also, conceptual, symbolic, linguistic, and technological intelligence this demonstrates.

There is clear evidence that UAP often are reported near nuclear sites. This link is documented unambiguously in the NDAA, which is written on the basis of both classified and unclassified information. Section 1683 of the NDAA details the link between UAP and nuclear technologies clearly in (h) ANNUAL REPORT, (2) ELEMENTS. This section of the NDAA states that each annual report on UAP to the appropriate congressional committees should include:

> (M)   The number of reported incidents, and descriptions thereof, of unidentified aerial phenomena associated with military nuclear assets, including strategic nuclear weapons and nuclear-powered ships and submarines.

> (N)   In consultation with the Administrator for Nuclear Security, the number of reported incidents, and descriptions thereof, of unidentified aerial phenomena associated with facilities or assets associated with the production, transportation, or storage of nuclear weapons or components thereof.

> (O)   In consultation with the Chairman of the Nuclear

Regulatory Commission, the number of reported incidents, and descriptions thereof, of unidentified aerial phenomena or drones of unknown origin associated with nuclear power generating stations, nuclear fuel storage sites, or other sites or facilities regulated by the Nuclear Regulatory Commission (NDAA 2021, 581-82).

To summarize, the NDAA clearly states that UAP activity is correlated with nuclear weapons sites; nuclear-powered ships and submarines; facilities associated with the production, transport, and storage of nuclear weapons; nuclear power generating stations; and nuclear fuel storage sites.

Many people erroneously gloss the recent history of UAP sightings by stating that UAP activity significantly began to be noticed in the 1940s. This is incorrect. It was the prior decade of the 1930s when UAP activity really got off the ground (pardon the pun). In fact, recent UAP waves were reported even earlier, in 1896 and in 1909. UAP activity really began to pick up in the 1930s, however, primarily in Scandinavia and also in the U.S. and U.K. (Keel 1970-71; see also Good 1987) when so-called ghostfliers — reported sporadically in 1932 and 1933—began to create quite a stir in December 1933 (Keel 1970-71, Part 1, 10). This was around the time that human beings first were learning about radioactivity. For example, in 1934, physicist Enrico Fermi and his team in Rome were bombarding various elements with neutrons to create radioactive elements. Even though they did not know it at the time, they split the uranium atom (Schwartz 2019).

In other words, people began noticing UAP with considerable frequency just as research was underway that would culminate not long thereafter with the development in the 1940s of the atomic bomb. Hitler's reliance on Sweden's iron ore as a key element of his military armaments (Karlborn 1965) also may be part of the reason that UAP sightings were so prevalent in Scandinavia starting at least in the 1930s.

The fact that the connection between UAP sightings and nuclear sites goes back decades is unsurprising given the risk that nuclear weapons pose to humankind. My view is that ETI was prescient in assessing the risks to human society even before human beings became capable of splitting the atom but while they still were in the initial stages of this research.

Extending the historical narrative relating to UAP sightings to include the decade of the 1930s raises a rather fantastical yet entirely plausible idea, one that dovetails with this book's emphasis on the inseparability of consciousness and matter. The idea is simple—namely that the ETI keeping tabs on humankind is not merely interested in physical events, such as nuclear explosions, but it also is interested in human consciousness, as reflected in human *motivations and human ideology*—and that when people with despotic motivations artificially create conflicting ideologies and then back these conflicting ideologies with devastating weapons systems and actual violence, ETI makes its presence more clearly known.

In other words, the relationship between ETI and humankind is far more nuanced that ETI's wise disdain for and significant concern regarding all things nuclear. It is not nuclear technologies alone, i.e., the physical side of things, or negative mental patterns alone, i.e., the consciousness side of things, that catalyzes ETI to act. Because ETI understands the inseparability of consciousness and matter, it evaluates how human thinking and human technological capabilities are interwoven and sculpts its interactions with human beings accordingly. ETI's understanding of the inseparability of consciousness and matter means that ETI's inherent nature requires it to act as a stabilizing presence vis-à-vis the whole. My view is that many UAP have been seen and otherwise detected in proximity to nuclear weapons storage facilities and other nuclear installations from the 1940s to the present (Tritten 2021; Aerial Phenomena 2021c) precisely because nuclear weapons are so destabilizing to the whole.

In 1948, UAP sightings occurred near military installations in the U.S., including the Los Alamos and Sandia atomic weapons laboratories in New Mexico. UAP reported there include relatively defined discs and saucers and numerous, amorphous green fireballs (King 2020 <2018>). Formerly classified USG documents released in accordance with the Freedom of Information Act (FOIA) confirm that as early as December 1948, UAP described as discs and saucers began to be sighted in Los Alamos, in proximity to the Atomic Energy Commission Installation. A little more than three years earlier, this laboratory tested the first atomic bomb in the desert near Alamogordo, New Mexico. Documents state that throughout the 1950s, as many as 150 observations were made of such objects in Los Alamos and in proximity to the Sandia National Laboratories installation in nearby Albuquerque, New Mexico. Other declassified documents discuss UAP sightings at the Oak Ridge National Laboratory in Tennessee [currently managed by UT-Battelle for the U.S. DoE], the Hanford Site plutonium processing plant in Washington State, and the Savannah River Site, another plutonium manufacturing site that is located in South Carolina (Hastings 2016).

Built in 1943 and now decommissioned, the Hanford Site was the first plutonium production plant in recorded history. UAP were seen there regularly. From 1947 onward, many UAP sightings also occurred at Oak Ridge. For example, an FBI Office Memorandum dated January 31, 1949, to the Director of the FBI, who at that time was Hoover, describes an event at Oak Ridge in June 1947 in which an unidentified aerial object was observed visually and tracked on radar. This was only one month prior to the famous UAP event in Roswell, New Mexico. Although the object at Oak Ridge was photographed and multiple copies of this photograph were made, this important photograph never was released to the public (Aerial Phenomena 2021c).

From 1948 to 1950, UAP sightings occurred so regularly at Oak Ridge that FBI memos listed dozens of such incidents in chronological order. One interesting sighting of an unidentified aerial object occurred at approximately 3:00 p.m. on July 19, 1953 and lasted around five minutes. A memo describes the unidentified aerial object as "extremely black in color, having the appearance, of a deep black metal exterior, with a fine gloss." The memo also states that the object did not leave a vapor trail and that no sounds were heard. According to the memo, witnesses reported that the object flew east and stopped before being "joined, by two more of these same objects. A formation, similar to a spread V was formed, and the objects at a tremendous speed flew in an eastward direction" (Aerial Phenomena 2021c).

The aforementioned January 31, 1949 FBI Office Memorandum describes numerous UAP detected in 1948 and 1949 in New Mexico. Multiple, relatively amorphous, glowing fireballs were first reported on December 5, 1948, with reports continuing through January 1949. Fireballs were detected in the vicinity of the Atomic Energy Commission (AEC) installation and Los Alamos National Laboratory, both in Los Alamos, New Mexico. The colors of these fireballs were described variously, as greenish white, bright green, yellow green, or blue green. In addition, a commercial airline pilot described an orange fireball that turned green. Observers noted that the fireballs followed a horizontal path, sometimes with slight variations, and that they travelled at speeds less than that of meteors but more than that of any type of known aircraft. The UAP were visible from one to five seconds, no sound was discernible by witnesses, and no trails or dust clouds were detected. The Association of Los Alamos Scientists (ALAS), the members of which all held security clearances, concluded that the objects were not meteors, but the group did not determine what the UAP actually were (Aerial Phenomena 2021b). Interestingly, ALAS was founded on August 30, 1945, and its membership included scientists who had worked on the development of the atomic bomb (University of Chicago Library 2009).

The UAP sightings of December 1948 and January 1949 were of such importance that USA and USAF Intelligence were alerted to the existence of the fireballs and the topic was deemed Top Secret. Following these events, a USAF Office of Special Investigations (OSI) memorandum of October 1949 states that the U.S. military attempted to detect UAP during what was called Project Twinkle. A triangulation station comprised of three theodolite cameras placed in different regions of New Mexico was used, and despite methodological shortcomings, Project Twinkle did detect one unidentified aerial object of approximately thirty feet in diameter traveling multiple thousands of miles per hour at an elevation of around 100,000 feet (Aerial Phenomena 2021b).

In late 1949, a significant UAP event occurred involving "a key atomic base" that most likely was located in New Mexico. The incident involved a very high-ranking USAF officer and five metallic UAP tracked on radar travelling at approximately 4,500 miles per hour at an elevation of around 100,000 feet. During a similar incident, another unidentified aerial phenomenon was tracked traveling at approximately the same speed at around the same altitude (Aerial Phenomena 2021b).

In March 1952, a UAP event specifically associated with uranium mines occurred in the southern part of the Belgian Congo. Multiple UAP described as "fiery discs flying in elegant curves" were sighted over active uranium mines in the Elisabethville District. The event took place to the east of the Luapula River, which connects the Meru and Bangweolo lakes. Comments from Commander Pierre, a fighter pilot who approached within approximately 120 meters of one of the unidentified objects, were compiled in a report dated March 29, 1952. Commander Pierre described the object as disc shaped, with a color similar to aluminum and with a diameter of between twelve and fifteen meters (Aerial Phenomena 2021c).

Multiple witnesses from around the world have observed and otherwise detected UAP activity in proximity to nuclear weapons storage and launch sites and underground nuclear launchers and launch control facilities (Toftenes and Toftenes 2010). In some instances, the nuclear weapons systems appear to have been either disarmed or armed remotely (Hastings 2016). In the U.S., for example, UAP activity has been associated with multiple intercontinental ballistic missile (ICBM). UAP events at ICBM sites occurred at the F.E. (Francis E.) Warren Air Force Base (AFB) in July and August 1965; Minot AFB in August 1966; Malmstrom AFB in March 1967; Malmstrom AFB again in November 1975; Minot AFB again in November 1975; Loring AFB in October and November 1975; and Wurtsmith AFB in October 1975. The latter two events involved Weapons Storage Areas (WSA) (UFOs & Nukes n.d.; see also Huyghe 1979). Informal reports also exist of such activity occurring in other countries such as the former USSR and, more recently, Russia.

Many people who witness UAP around nuclear installations come away with the impression that ETI is sending a message to humankind that human beings should not possess let alone use nuclear weapons and other nuclear technologies (Ridge 2015 <2003>; Hastings 2016). As long as humankind continues to possess nuclear weapons, it is at risk of annihilating itself. This risk increases exponentially if human beings can hack the launch and deactivation protocols for nuclear weapons systems from a distance, since any individual or group with this ability could extinguish the human species and much else with it. My view is that ETI is sending human beings a sublimely succinct message by arming and disarming humankind's nuclear weapons—***nuclear weapons systems are in fact hackable from a distance. You therefore must abolish all nuclear weapons on Earth, and you should not position nuclear weapons in space***.

Considerable evidence supports the conclusion that ETI does not want any weapons—particularly nuclear weapons—in space. In 1964, an unidentified aerial object fired multiple beams of light and shot down a nuclear missile with dummy warheads after it was launched at Vandenberg Air Force Base in California (Hastings 2016). Accordingly, if we want to forge constructive communication with ETI, humankind clearly must demilitarize and de-weaponize space and must stop the arms race on Earth. Tragically, human beings appear to be going in precisely the wrong direction. Within Earth's atmosphere, for example, the U.S., China, and Russia are competing fiercely with respect to hypersonic weapons (Andresen 2021).

The international security situation with respect to space also is deteriorating rapidly, with the space race making humankind's possession of nuclear weapons increasing problematic (Mecklin 2022). In response to the high degree of competition between the U.S. and China, which includes conflict with respect to Taiwan and the South China Sea, NASA is advocating for the use of nuclear-powered spacecraft to keep the U.S. ahead of China in the space race (Gohd 2021). I understand that NASA may be quite displeased that China is basing its current development of a hypersonic nuclear missile engine that can travel at 6,000 miles per hour on a former NASA design (Knox 2021). I also understand that the USG more generally may be quite frustrated that China has copied U.S. nuclear weapons technology in the past, and, even worse, passed its nuclear weapons technology to North Korea in a clear attempt to try to destabilize the U.S. Nevertheless, my strong position here is that nuclear-powered spacecraft are a very bad idea. We are smart enough as a species to figure out another way to manage ourselves geopolitically and to further exploration of space without resorting to putting nuclear technologies in space.

In the last two decades, reports indicate that UAP activity either has increased around U.S. military sites, and/or that it is being detected more often now because the military is using new sensor systems. As mentioned above, Administrator Bill Nelson of NASA recently stated that USN pilots have reported more than three hundred UAP sightings since 2004 (Nelson 2021). In November 2004, well-documented UAP activity occurred around the USS *Nimitz* (Beaty 2019 <2018>), which is a nuclear-powered carrier strike group (Pike 2000; Wikipedia 2020). In 2015, UAP activity occurred around the USS *Theodore Roosevelt* (History n.d.), which also is a nuclear- powered carrier strike group (Wikipedia 2021).

UAP activity around nuclear-powered carrier strike groups indicates that ETI is concerned about potentially hostile actions by these vessels and, also, that accidents could occur involving the nuclear technologies propelling these vessels. In fact, nuclear-powered submarines have a long history of accidents. Two U.S. nuclear submarines, USS *Thresher* and USS *Scorpion*, currently are sitting at the bottom of the Atlantic Ocean where they sank in the 1960s (Keane 2021). More recently, in 2019, many senior Russian naval officers were killed in an accident involving one of Russia's deep- submergence vehicles (DSV). It is believed that DSVs may have been conducting covert sea trials of weapons system comprised of a large caliber nuclear torpedo (Muraviev 2019).

In late March and early April of 2019, UAP were detected near the U.S. Army's Terminal High Altitude Area Defense (THAAD) battery near the North West Field at Andersen Air Force Base on the Island of Guam. Guam is strategically important in the context of geopolitical dynamics involving China and North Korea, and the THAAD battery is tasked with defending the island from ballistic missile attacks (Rogoway and Trevithick 2020a).

In addition to the aforementioned events, beginning on July 14, 2019 and continuing over a number of days, groups of UAP followed USN vessels in the area around California's Channel Islands. The events centered on the *Arleigh Burke* class destroyer USS *Kidd*, the nearby USS *Rafael Peralta*, and the USS *Russell*. As many as six UAP are reported to have swarmed the ships at one time. In the case of the USS *Rafael Peralta*, a white light hovered over the ship's flight deck for more than ninety minutes, which is much longer than the twenty-eight minute maximum flight time of a Phantom IV drone, for example. The UAP matched the ship's speed of sixteen knots as it continued to hover over the vessel's helicopter landing pad. For almost three hours on July 15, 2019, significant UAP activity also occurred around the USS *Russell*, between San Clemente Island and San Diego but closer to shore as compared to the UAP activity that had occurred on July 14, 2019 (Kehoe and Cecotti 2021).

In addition to events around military bases and military vessels, many UAP sightings also have occurred at nuclear power generating stations. Two recent series of events occurred in 2019 and 2022, in the U.S. and Sweden, respectively. Whereas the media describes unidentified aerial phenomena near nuclear weapons facilities and nuclear-powered vessels as "UAP," it uses the term "drones" to describe the objects involved in these two series of events. The first series of events began on September 29, 2019 and continued on successive nights. During this period, a swarm of drones appeared at the Palo Verde Nuclear Generating Station near Tonopah, Arizona. Palo Verde is the most powerful nuclear plant in the U.S. (Rogoway and Trevithick 2020b). This is not an isolated event, either. Nuclear Regulatory Commission (NRC) documents report that between 2015 and 2019, as many as sixty different drone sightings occurred at twenty-four nuclear power facilities in the U.S. (Tingley 2022; Trevithick 2022).

The second abovementioned series of events occurred in Sweden, where drone sightings were reported over active nuclear power facilities in Forsmark, Oskarshamn, and Ringhals (Mossige-Norheim and Strömberg 2022). According to the Swedish Police Authority, on January 14, 2022, drones were seen over Forsmark, which is the largest single energy- producing facility in Sweden and, also, over Oskarshamn (Tingley 2022). Drones also were reported at Ringhals and over the Barsebäck Nuclear Power Plant. The Barsebäck site was decommissioned in 2015 and is scheduled to be demolished by 2028. Furthermore, drones were reported over government buildings in the Swedish capital of Stockholm, such as Stockholm Palace, the official residence of Sweden's King Carl XVI Gustaf. One unmanned aircraft also was reported over the Swedish Riksdag, or legislature. In addition, drones were reported near Kiruna Airport (which is north of the polar circle) and near Luleå Airport (in proximity to the Gulf of Bothnia, which separates Sweden from Finland) (Trevithick 2022).

It is interesting to consider why the media uses the acronym "UAP" in many instances involving unidentified aerial phenomena but instead uses the term "drones" in the two aforementioned series of events and in other instances involving unidentified objects flying over nuclear power generating stations. Returning to the NDAA, of the three sections explicitly attesting to the correlation between UAP activity and nuclear sites —namely (h)(2)(M)-(O), as cited above—only (O) adds "or drones of unknown origin" after "unidentified aerial phenomena" (NDAA 2021, 581-82). To me, the language of the NDAA indicates that authorities are aware that at least in the U.S., the UAP described in (M) and (N) *are not* of terrestrial origin, and, further, that authorities also are aware that *both* UAP of unknown origin (i.e., extraterrestrial UAP) *and* drones of terrestrial origin are flying over nuclear power plants. I think that both types of UAP—human and extraterrestrial—also are operating in Sweden.

My view is that the human beings and ETI operating UAP over nuclear power plants are doing so for completely different reasons. Hostile human actors would be operating drones near strategic sites such as nuclear power plants to gather intelligence in order to plan attacks that could damage and/or destroy both military and civilian targets, commercial nuclear power infrastructure, and other elements of the power grid. Collecting information on wireless network cybersecurity at these sites assists hostile actors identify and assess physical and/or other vulnerabilities (Trevithick 2022). More specifically, by mapping wireless networks to detect cybersecurity vulnerabilities relating to critical infrastructure, hostile actors can collect information they plan to use later during weaponized drone attacks intended to sabotage utility grids and nuclear infrastructure (Tingley 2022). In contrast, ETI may be operating UAP over nuclear power plants to ensure that it can mitigate accidents, environmental catastrophes, and/or other vulnerabilities in the event of destructive actions by human actors.

In addition to occurring frequently at nuclear sites, UAP activity also has been reported for years in the vicinity of volcanoes. For example, a recent video shows a UAP at the La Palma volcano at Cumbre Vieja (Old Summit) on the island of Tenerife in Spain's Canary Islands (iceage2012 2021). Ash from volcanic eruptions impacts Earth in devastating ways. The possibility that a volcanic eruption could damage or even destroy a nuclear reactor is a horrific, potential risk. The idea that ETI is preparing to mitigate such a disaster if and when it may occur connects ETI's concern with nuclear issues to UAP activity near volcanoes.

In fact, a scenario involving a volcanic eruption destroying a nuclear reactor cannot be ruled out anywhere. For example, certain areas of the world such as Japan have numerous active volcanoes. Around 110 of the approximately 1,500 active volcanoes in the world are located in the so- called Ring of Fire, which includes Japan within it (Jozuka 2015). Nevertheless, nine nuclear reactors were operating in Japan as of March 2021, even though Japan is a relatively small country territorially. At one of these facilities, the Kashiwazaki-Kariwa nuclear power plant, an employee identification card was used in an unauthorized manner in September 2020 to gain access to the central control room (nippon 2021).

Now, almost ninety years since ETI brought itself to humankind's attention by means of its ghostfliers, and *thanks to ETI*, human beings are coming out of denial about the existence of ETI at precisely the same time that they urgently must come out of denial with respect to the seriousness of the nuclear issue on Earth and in space. This is occurring because ETI's UAP are appearing around nuclear weapons facilities, nuclear-powered ships and submarines, nuclear power plants, etc., thereby managing to convey two messages simultaneously—yes, ETI exists, and please, human beings, do something to abolish nuclear weapons and other nuclear technologies. The situation on Earth is precarious enough to begin with, and there is no need to amplify this precariousness by permitting certain human beings to possess and control nuclear weapons and other nuclear technologies.

The heartening conclusion we can draw with respect to the correlation between extraterrestrial UAP events and nuclear sites is that ETI is very attuned to the underlying wholeness of reality. ETI is light-years ahead of humans in understanding the extent to which nuclear detonations and other destructive activities rip through the whole. By means of its repeated presence at nuclear sites, ETI is signaling very clearly to human beings that nuclear weapons are unacceptable. This is a very wise position. Given the extreme perilousness of the entire nuclear situation, humankind should heed ETI's lucid warning and take all necessary steps to abolish nuclear weapons and other nuclear technologies on Earth and in space.

CHAPTER 6

Moving Beyond Fragmentation and Reductionism

The ETI around us now realizes that the entire nuclear paradigm constitutes misinformation a central position in human society. Tragically, the scientific research that led to the first test of an atomic weapon in recent history on July 16, 1945 was mirrored by the fragmentation of human consciousness. As the nuclear paradigm has continued to expand on Earth since WWII, physical fragmentation continues to be mirrored by cultural fragmentation within society, and, also, by fragmentation within humankind's cartographies of knowledge and its map of disciplines in the academy.

Humankind's nuclear predicament is not merely physical, either. Paradigmatically, a framework based on fission represents a fragmented and partial view that fails to perceive how all aspects of reality are a whole. The concept of fragmentation describes nuclear fission precisely, since fission literally splits the nucleus of the atom. In contrast, the ETI around us here on Earth does not differentiate the mental and physical aspects of events, situations, and historical contexts —the mental and physical aspects of everything are perceived as inseparable for this ETI, who recognizes that splitting physical atoms during nuclear fission also *inherently* fragments consciousness. Given the extent to which first-generation atomic bombs and second-generation thermonuclear weapons, fusion weapons, and hydrogen bombs (H bombs) tear through the fabric of existence and disrupt the whole, humankind's use of these weapons must look like the worst kind of barbarism to this sophisticated ETI who values coherence in consciousness, peacefulness in society, and tranquility in the Cosmos.

Because it is based on an incorrect dichotomy between consciousness and matter, the cartography of human knowledge reflected in the current map of academic disciplines is obsolete. Fundamental misinformation such as consciousness/matter dualism and the idea that nothing exists beyond explicate order occupies central positions within human society. This causes confusion and chaos at every turn. Academia does not escape this chaos, which appears methodologically in all current disciplines in a manner that mirrors the nature of the misinformation itself.

Clear perception of the nature of reality requires a *whole*scale (Bohmian pun intended) overturning of it all. This includes the current map of academic disciplines, the organizational structure of academic institutions, and, as I discuss in an interview (Kalantarova 2021), the multiplicity of fragmented methodologies that direct academic scholarship and research. Fragmented methodologies never move beyond an explicate order perspective to achieve a deeper understanding of implicate order, holomovement, and many other important concepts.

With respect to academic disciplines and groups of disciplines, how will perception of the inseparability of consciousness and matter and awareness of wholeness impact major categories of knowledge, such as religion and science? Here, we should strive to learn from the ETI around us now. For this advanced ETI, the reduction of mind toward idealism and the reduction of matter toward materialism do not occur. This means that categorical distinctions such as science and religion that emanate from the false dichotomy between consciousness and matter also are not meaningful for this ETI.

Any ETI that perceives the inseparability of consciousness and matter does not perceive reality according to a divide between so-called objective science and so-called subjective religion and theology. Instead, such an ETI recognizes that both science and religion share the same ontological ground. Human beings have much to learn from this understanding, particularly since materialist interpretations in science and ungrounded interpretations in theology both are commonplace. These extremes should be avoided, since they are based on mistaken perception of and misinformation regarding the nature of reality itself.

During a panel discussion at the Stedelijk Museum in Amsterdam (Art Meets Science and Spirituality 1990), Bohm made a very insightful observation regarding the nature of science itself. A portion of this panel discussion is shown in the film *Infinite Potential*, which focuses on Bohm's life and career:

> Science is whatever people make of it. You see, science has changed over the ages—it's different now from a few hundred years ago, it will be different again. Now, there's no intrinsic reason why science must necessarily be measurement. This is another historical development which has come about over the past few centuries—it's entirely contingent, not absolutely necessary (Howard 2020).

Scientists who follow a methodology that relies only on measurement and data collected by means of instruments should consider that reality is more complex than what can be described in a reductionistic manner. In the case of ETI, it is valuable to expand one's methodology beyond mere measurement and open the door to the phenomena. Broadening one's methodology does not make a person unscientific, it merely reflects the fact that science itself is not fixed but must expand its own horizons to meet the wholeness of reality. Similarly, religion and theology also must expand to meet the fullness of reality that ETI is elucidating. At some point, then, science and other more qualitative disciplines such as religion and theology will find themselves meeting as they explore the fullness of existence together.

The ETI around us here on Earth seems particularly attuned to humankind's need to move beyond the current fragmentation of academic disciplines, and, indeed, to move beyond fragmentation more generally. To further creative acculturation with this ETI, human beings therefore must deepen their realization of the inseparability of consciousness and matter and they must reason deductively from the perspective of the whole. Starting from direct experience of the wholeness of reality and the preciousness of existence, human action will become more centered and more coherent. This will facilitate a relational perspective that recognizes and nurtures a creative and positive relationship with ETI.

In anticipation of widespread Contact—capitalized here to indicate Contact over essentially the entire human species for the purpose of differentiating this from individual instances of contact, which many people state already have occurred — it is imperative for each individual human being to assume an internal locus of control stance whereby one embodies the recognition that events in human history result from one's own actions. Unlike external locus of control thinking by means of which one capitulates agency to an imagined 'other,' taking an internal locus of control approach means that we proceed in a deliberate and psychologically centered manner by realizing that we ourselves determine the implications for human society of ETI's existence around us. While I am certain that ETI's intentionality towards humankind is kind and magnanimous, this is not what determines the implications of widespread Contact for human society. We do.

For now, humankind should focus its efforts on communicating constructively and creatively with the ETI around us here on Earth, and with any ETIs we encounter in the future. In keeping with Ambassador Woolsey's aspirational hope that humankind can be friends with ETI, quoted above, I am certain that we can be friends with the ETI in our midst now and with any other ETIs that come our way. It would be a very significant and helpful gesture in the developing relationship between humankind and ETI if humankind would abolish nuclear weapons and eliminate other nuclear technologies. As humankind increases in wisdom, it will eliminate weapons altogether, and all the resources wasted on weapons design, manufacturing, testing, and deployment instead will be used to foster constructive human creativity and a compassionate global society that embodies principles such as kindness, social justice, and love. We can achieve this as a species—and, as we do, we will find ourselves developing a closer relationship with the ETI around us now and other ETIs we may be fortunate enough to encounter in the future.

REFERENCES

Aerial Phenomena. 2021a. "The CIA's UFO Admission." Season 2, Episode 1 of *Aerial Phenomena. Gaia.* Premiered October 10, 2021.

—. 2022. "The Curious Case of the CIA Director Roscoe Hillenkoetter." Season 2, Episode 4 of *Aerial Phenomena. Gaia.* Premiered January 30, 2022.

—. 2021b. "What Were the Green Fireballs?" Season 1, Episode 7 of *Aerial Phenomena. Gaia.* Premiered September 23, 2021.

—. 2021c. "Why Are UFOs/UAPs Interested in Nuclear Technology?" Season 1, Episode 2 of *Aerial Phenomena. Gaia.* Premiered September 30, 2021.

—. 2021d. "The Year UFO Secrecy Nearly Ended!" Season 2, Episode 2 of *Aerial Phenomena. Gaia.* Premiered October 10, 2021.

Agreement (Agreement on Measures to Reduce the Risk of Outbreak of Nuclear War Between the United States of America and the Union of Soviet Socialists Republics – September 30, 1971). 1971. The Avalon Project, Yale Law School Lillian Goldman Law Library. https://avalon.law.yale.edu/20th_century/sov001.asp.

Andresen, Jensine. 2023. *Extraterrestrial Ethics.* London: Ethics International Press.

—. In preparation. *Extraterrestrial Mind.*

—. 2021. "Two Elephants in the Room of Astrobiology." In *Astrobiology: Science, Ethics, and Public Policy*, edited by Octavio A. Chon Torres, Ted Peters, Joseph Seckbach, and Richard Gordon, 193-231. Hoboken, NJ/Beverly MA, Wiley/Scrivener.

ArtisMortisGallery. 2021. "UFO Identification Chart 1967." *Etsy.com*, December 22, 2021. Accessed December 26, 2021. https://www.etsy.com/listing/944646837/ufo-identification-chart-1967- flying.

Art Meets Science and Spirituality in a changing Economy. 1990. Conference, Stedelijk Museum, Amsterdam, September 14-19, 1990. https://www.stedelijk.nl/en/collection/33917-willem-velthoven-art-meets-science-and-spirituality-in-a-changing-economy. [In 1996, presentations from this even were compiled and published in book form together with additional presentations given at a second event that occurred in Denmark in 1996. In 2015, Mystic Fire Video created a video of the 1990.]

Beaty, David C., director. 2019 <2018>. *The Nimitz Encounters.* Uploaded May 26, 2019. [The film previously was uploaded in 2018.] https://www.youtube.com/watch?v=PRgoisHRmUE.

Bennett, Jay. 2020. "Here's How Much Deadlier Today's Nukes Are Compared to WWII A-Bombs." *Popular Mechanics*, December 13, 2020. https://www.popularmechanics.com/military/a23306/nuclear- bombs-powerful-today/.

Best, Jr., Richard A. 2004. *Intelligence Community Reorganization: Potential Effects on DOD Intelligence Agencies.* CRS [Congressional Research Service] Report for Congress, December 21, 2004. https://irp.fas.org/crs/RL32515.pdf. [Page numbers cited are from the PDF.]

—. 2010. *Intelligence Reform After Five Years: The Role of the Director of National Intelligence (DNI).* CRS [Congressional Research Service]. CRS Report for Congress, June 22, 2010. https://sgp.fas.org/crs/intel/R41295.pdf.

Bohm, David. 1987. "Hidden variables and the implicate order." In *Quantum Implications: Essays in Honour of David Bohm*, edited by Basil J. Hiley and F. David Peat, 33-45. Abingdon, U.K.: Routledge.

—. 1986. "The implicate order and the super-implicate order." In *Dialogues with Scientists and Sages: The Search for Unity*, edited by Renée Weber, 23-49. London: Routledge & Kegan Paul.

—. 1989. "Meaning and Information." In *The Search for Meaning: The New Spirit in Science and Philosophy*, edited by Paavo Pylkkänen, 43-62. Wellingborough, UK: Crucible. https://

www.implicity.org/Downloads/Bohm_meaning
+information.pdf. [Page numbers cited are from the PDF.]

—. 1990. "A new theory of the relationship of mind and matter." *Philosophical Psychology* 3, no. 2: 271-86.

—. 1952. "A suggested interpretation of the quantum theory in terms of hidden variables." *Physical Review* 85: 166-89.

—. 1980. *Wholeness and the Implicate Order.* Abingdon, U.K.: Routledge & Kegan Paul.

Bohm, David, and Basil J. Hiley. 1987. "An ontological basis for the quantum theory." *Physics Reports* 144: 323-48.

—. 1995. *The Undivided Universe: An Ontological Interpretation of Quantum Theory.* New York: Routledge.

Brennan, John O. 2021. "The Global Beacon – Special Guest Ex CIA Director John O. Brennan." *Global Beacon*, June 24, 2021. https://podcasts.google.com/feed/
aHR0cHM6Ly9mZWVkcy5idXp6c3
Byb3V0LmNvbS8xNTU4MzI4LnJzcw/episode/
QnV6enNwcm91dC0   4NzU5MTc4?
sa=X&ved=0CAUQkfYCahcKEwjYh9yL8Iv2AhUAA
AAAHQAAAAAQAg&hl=en.

—. 2020. "John O. Brennan on Life in the CIA: What working in intelligence has taught him about human nature." Episode 111 of *Conversations with Tyler T. Cowen.* December 16, 2020. https://conversationswithtyler.com/
episodes/john-o-brennan/.   [Mercatus Center uploaded the video version of this interview, which is available at https://
www.youtube.com/watch?v=LdQ7L0ugZJc.]

CNN   Editorial   Research   (CNN).   2021.   "CIA Directors   Fast   Facts."   *CNN*,   September   24,   2021. https://www.cnn.com/2013/11/12/us/cia-directors-   fast-facts/index.html.

COBEPS (Comité belge d'étude des phénomènes spatiaux). 2021. "Catalogue des Observations Belges: 979 - août 1989." https://datastudio.google.com/u/0/
reporting/7e5a715a-212b-473d-   8491-992f94b87605/
page/6kSMC?s=l6cZ8UAgv8U.

Cohen, Zachary. 2021. "US intel and satellite images show Saudi Arabia is now building its own ballistic missiles with help of China." *CNN*, December 23, 2021. https://www.cnn.com/2021/12/23/politics/saudi-ballistic-missiles-china/index.html.

Cordesman, Anthony H. 2005. *Proliferation of Weapons of Mass Destruction in the Middle East: The Impact on The Regional Military Balance*. Working Draft: Revised March 25, 2005. Washington, D.C.: Center for Strategic and International Studies (CSIS). https://csis-website-prod.s3.amazonaws.com/s3fs-public/ legacy_files/ files/media/csis/pubs/050325_proliferation %5B1%5D.pdf. [Page numbers cited are from the PDF.]

Creighton, Jolene. 2014. "The Kardashev Scale – Type I, II, III, IV & V Civilization." *Futurism*, July 19, 2014. https://futurism.com/.

DeVine, Michael E. 2021. "Defense Primer: Under Secretary of Defense for Intelligence and Security." *Congressional Research Service* (CRS), December 13, 2021. https://sgp.fas.org/crs/natsec/IF10523.pdf.

FAS (Federation of American Scientists). n.d. *The External Referral Working Group: Its Origins and Development*. Project on Government Secrecy. https://sgp.fas.org/advisory/erwg.html.

FBI (Federal Bureau of Investigation). 1950. "Guy Hottel Part 1 of 1." *FBI Records: The Vault*. https://vault.fbi.gov/hottel_guy/Guy %20Hottel%20Part%201%20of%2 01/view.

—. 2013. "UFOs And The Guy Hottel Memo." *FBI.gov*, March 25, 2013. https://www.fbi.gov/news/stories/ufos-and-the-guy-hottel-memo.

Glassel, Roger. 2021. "NEW: The UAP activities in NDAA 2022 will be implemented as activities to the Airborne Object Identification & Management Synchronization Group (AOIMSG). Not a new office. #ufo #uap #ufotwitter #uaptwitter #ndaa2022." *Twitter*, December 29, 2021. https://twitter.com/rogerglassel/

status/1476121279266705410?s=20. [This tweet to Mr. Roger Glassel includes a screenshot of an email from Colonel Sue Gough, who is retired from the Army and who currently works as Senior Strategic Planner and Pentagon spokesperson.]

Gohd, Chelsea. 2021. "NASA thinks US needs nuclear-powered spacecraft to stay ahead of China." *Space.com*, October 23, 2021. https://www.space.com/us-needs-nuclear-powered-spacecraft.

Good, Timothy. 1987. *Above Top Secret: The worldwide UFO cover-up*. London: Sidgwick & Jackson Ltd.

Gross, Judah Ari. 2021. "Israel receives 3 more F-35 fighter jets." *The Times of Israel*, April 25, 2021. https://www.timesofisrael.com/liveblog_entry/israel-receives-3-more- f-35-fighter-jets/.

Haines, Avril. 2021. "Our Future in Space: Ignatius Forum." Washington National Cathedral, November 10, 2021. https://www.youtube.com/watch?v=UWyPk_f8aAA.

Haines, Gerald K. 1997. "The CIA's Role in the Study of UFOs, 1947-90." *Studies in Intelligence* 1, no. 1: 67-84.

Hastings, Robert, director. 2016. *UFOs and Nukes: The Secret Link Revealed*. Verifiable Pictures. [Topics discussed in this film also are discussed in Hastings, Robert, 2017 <2008>, *UFOs & Nukes: Extraordinary Encounters at Nuclear Weapons Sites*. Second edition, revised and updated. No publisher is listed for the second edition, but the first edition was published in 2008 by AuthorHouse.]

Hicks, Kathleen H. 2021. "Establishment of the Airborne Object Identification and Management Synchronization Group." Memorandum for Senior Pentagon Leadership, Commanders of the Combatant Commands, Defense Agency and Field Activity Directors. United States Department of Defense, November 23, 2021. https://www.airforcemag.com/app/uploads/2021/11/ESTABLISHMEN   T-OF-THE-AIRBORNE-OBJECT-

IDENTIFICATION-AND- MANAGEMENT-
SYNCHRONIZATION-GROUP.pdf.

History.com (History). n.d. "USS Roosevelt 'Gimbal'
UFO: Declassified Video." https://www.history.com/topics/
paranormal/uss-roosevelt-gimbal- ufo-declassified-video.

Howard, Paul, director. 2020. *Infinite Potential: The Life
& Ideas of David Bohm.* Imagine Films. https://
www.infinitepotential.com/.

Huyghe, Patrick. 1979. "U.F.O. Files: The Untold
Story." *The New York Times*, October 14, 1979,
106. https://www.nytimes.com/1979/10/14/archives/ufo-
files-the-untold- story.html.

IAEA (International Atomic Energy Agency Board of Governors).
2021. *Verification and monitoring in the Islamic Republic of Iran
in light of United Nations Security Council resolution 2231
(2015).* Report by the Director General, September 7, 2021,
derestricted September 15, 2021. https://www.iaea.org/sites/
default/files/21/09/gov2021-39.pdf.

iceage2012. *UAP at Spain's La Palma volcano.* October 11, 2021.
https://imgur.com/gallery/cXeraw8.

IPFM (International Panel on Fissile Materials). 2021. "Fissile
material stocks." International Panel on Fissile Materials,
September 4, 2021. https://fissilematerials.org/.

—. 2015. *Global Fissile Material Report 2015: Nuclear Weapon
and Fissile Material Stockpiles and Production.* Eighth annual
report of the International Panel on Fissile Materials. https://
fissilematerials.org/library/gfmr15.pdf.

Iran International. 2021. "Iran Can Have Enough Uranium For A
Nuke In One Month – Report." *Iran International*, September
14, 2021. https://old.iranintl.com/en/world/iran-can-have-
enough-uranium-nuke- one-month-report.

Jackson, Herb. 2009. "North Bergen man is homeland security
assistant for President Obama." *NorthJersey.com*, December 5,
2009. https://web.archive.org/web/20110507023129/http://
www.northjersey. com/
news/120509_North_Bergen_man_is_homeland_security_ass

ista nt_for_President_Obama.html.

Jozuka, Emiko. 2015. "How Volcanic Eruptions Threaten Nuclear Power Plants." *Vice*, November 14, 2015. https://www.vice.com/en/article/pgaj9b/how-volcanic-eruptions- threaten-nuclear-power-plants.

Kalantarova, Olena. 2021. Methodological Pluralism Through the Lens of the Buddhist Doctrine of Time Kālacakra: An Interview with Dr. Jensine Andresen." *Філософська думка (Philosophical Thought)* 2: 165-83.

Karlbom, Rolf. 1965. "Sweden's iron ore exports to Germany, 1933-1944." *Scandinavian Economic History Review* 13, no. 1: 65-93. https://www.tandfonline.com/doi/pdf/10.1080/03585522.1965.104143 65.

Keane, Daniel. 2021. "Nuclear-powered submarines have 'long history of accidents', Adelaide environmentalist warns." *ABC (Australian Broadcasting Corporation) News*, September 17, 2021. https://www.abc.net.au/news/2021-09-17/nuclear-submarines-prompt-environmental-and-conflict-concern/100470362.

Keel, John A. 1970-71. "Mystery Aeroplanes of the 1930s." *Flying Saucer Review*, a four-part article, Part 1: vol. 16, no. 3 (May/June 1970): 10-13); Part 2: vol. 16, no. 4 (July/August 1970): 9-14; Part III: vol. 17, no
4 (July/August 1971): 17-19; Part IV: vol. 17, no. 5 (September/October 1971): 20-22, 28.
https://www.scribd.com/document/21787053/Mystery-Aeroplanes-of- the-1930-s-by-John-A-Keel.

—. 1969. "The Principle of Transmogrification." *Flying Saucer Review* 15, no. 4 (August): 27-28, 31. https://www.scribd.com/document/385489728/THE-PRINCIPLE-OF- TRANSMOGRIFICATION-by-John-A-Keel.

Kehoe, Adam, and Marc Cecotti. 2021. "Multiple Destroyers Were Swarmed By Mysterious 'Drones' Off California Over Numerous Nights." *The Drive*, March 23, 2021. https://www.thedrive.com/the- war-zone/39913/multiple-destroyers-were-swarmed-by-mysterious- drones-off-california-over-numerous-nights.

Kimball, Daryl. 2018. "Fissile Material Cut-off Treaty (FMCT) at a Glance." Accessed April 10, 2021. https://www.armscontrol.org/factsheets/fmct.

King, Darryn. 2020 <2018>. "Why Mysterious Green Fireballs Worried the
U.S. Government in 1948." *History Stories*, August 17, 2018, updated January 16, 2020. https://www.history.com/news/ufos-green-fireballs- nuclear-facilities-new-mexico.

Knox, Patrick. 2021. "China developing 6,000 mph hypersonic nuke missile 'based on design' ditched by NASA." *New York Post*, December 10, 2021.
https://nypost.com/2021/12/10/china-developing-6000-mph- hypersonic-nuke-missile-based-on-design-ditched-by-nasa/?utm_source=NYPTwitter&utm_campaign=SocialFlow&utm_medium= SocialFlow.

Krepon, Michael. 2021. "How to Avoid Nuclear War." *War on the Rocks*, November 8, 2021. https://warontherocks.com/2021/11/how-to-avoid- nuclear-war/.

Kristensen, Hans M., and Matt Korda. 2022. "The Nuclear Notebook: Israeli nuclear weapons, 2021." *Bulletin of the Atomic Scientists* 78, Issue 1: Special issue: Conflict in space: 38-50, January 17, 2022. https://www.tandfonline.com/doi/full/10.1080/00963402.2021.2014239.

Kristensen, Hans M., et al. 2021. "10. World nuclear forces." In *SIPRI [Stockholm International Peace Research Institute] Yearbook 2021: Armaments, Disarmament and International Security*. Oxford University Press. https://sipri.org/sites/default/files/2021-06/yb21_10_ wnf_210613.pdf. [Page numbers cited are from the PDF.]

Lock, Samantha. 2022. *Ukraine crisis live* (blog), *The Guardian*, February 23, 2022. "Russia open to 'diplomacy', says Putin, amid fresh shelling in east." Blog post entry 1:50 a.m., "Putin ready for 'diplomatic solutions' but Russia's interests non-negotiable." https://www.theguardian.com/world/

live/2022/feb/23/ukraine-russia- news-crisis-latest-live-updates-putin-biden-europe-sanctions-russian- invasion-border-troops.

McDonald, James E. 1969. "Science in Default: Twenty-Two Years of Inadequate UFO Investigations." Presentation at the American Association for the Advancement of Science, 134th Meeting, General Symposium, Unidentified Flying Objects, Boston, December 27, 1969.

Mecklin, John. 2022. "Interview: Robert Latiff on the worsening international security situation in space." *Bulletin of the Atomic Scientists* 78, Issue 1:
Special issue: Conflict in space: 3-5, January 17, 2022. https://thebulletin.org/premium/2022-01/interview-rob-latiff-on-the- worsening-international-security-situation-in-space/.

Mossige-Norheim, Thea, and Lars-Olof Strömberg. 2022. "Polisen: Drönare övar kärnkraftverk i natt." *Expressen*, January 15, 2022. https://www.expressen.se/nyheter/uppgift-till-polisen-foremal-over- karnkraftverken/.

Muraviev, Alexey. 2019. "Russia's nuclear submarine disaster will test President Vladimir Putin and his navy." *ABC (Australian Broadcasting Corporation) News*, July 3, 2019. https://www.abc.net.au/news/2019- 07-03/russias-nuclear-submarine-disaster-test-vladimir-putin- navy/11274964.

NASIC (National Air and Space Intelligence Center). n.d. "National Air and Space Intelligence Center Heritage." https://www.nasic.af.mil/About- Us/Fact-Sheets/Article/611728/national-air-and-space-intelligence- center-heritage/.

NDAA (National Defense Authorization Act for Fiscal Year 2022). 2021. Public Law 117-81, 135 Statute 1541-2450. 50 U.S. Code § 3373 (135 Statute 2118- 23). Section 1683. ESTABLISHMENT OF OFFICE, ORGANIZATIONAL STRUCTURE, AND AUTHORITIES TO ADDRESS UNIDENTIFIED AERIAL PHENOMENA. December 27, 2021.

https://www.congress.gov/117/plaws/publ81/
PLAW-117publ81.pdf.

Nelson, Clarence William ("Bill") Nelson II. 2021. "Interview
with Bill Nelson October 19, 2021." University of Virginia
(UVA) Center for Politics. https://www.youtube.com/watch?
v=9hH1XEqKlTs.

nippon (nippon.com). 2021. "Japan's Nuclear Power
Plants in 2021." *nippon.com*, March 31, 2021. https://
www.nippon.com/en/japan-data/h00967/.

NTI (Nuclear Threat Initiative). 2020. "Proposed Fissile
Material (Cut-Off) Treaty (FMCT)." Accessed April 10, 2021.
https://www.nti.org/education-center/treaties-and-regimes/
proposed- fissile-material-cut-off-treaty/.

Obama, Barack Hussein II. 2021. "Reggie Watts to Barack Obama:
What's w/Dem Aliens?" The Late Show with James Corden.
https://www.youtube.com/watch?v=xp6Ph5iTIgc.

ODNI (Office of the Director of National Intelligence). 2021.
Preliminary Assessment: Unidentified Aerial Phenomena, June
25, 2021. McLean, VA: Office of the Director of National
Intelligence. https://www.dni.gov/files/ODNI/documents/
assessments/Prelimary- Assessment-UAP-20210625.pdf.

OSD (Office of the Secretary of Defense). 2021.
"Office of the Secretary of Defense." U.S. Department
of Defense. https://www.defense.gov/About/Office-of-the-
Secretary-of-Defense/.

OUSDI (Office of the Under Secretary of Defense for Intelligence
& Security). n.d. "Office of the Under Secretary of Defense for
Intelligence & Security." https://ousdi.defense.gov/.

Perkins, Richard H., Michelle T. Bensi, Jacob Philip, and Selim
Sancaktar. 2011. *Screening Analysis Report for the Proposed
Generic Issue on Flooding of Nuclear Power Plant Sites
Following Upstream Dam Failures*. U.S. Nuclear Regulatory
Commission, Office of Nuclear Regulatory Research, Division
of Risk Analysis. https://big.assets.huffingtonpost.com/
flooding.pdf.

Pike, John. 2000. "CVN-68 Nimitz-Class." FAS (Federation of

American Scientists) Military Analysis Network.
https://man.fas.org/dod-101/sys/ship/cvn-68.htm.

Polansky, Anne (with the assistance of Lawrence Criscione). 2018. "As Fossil Fuels Melt the Planet, Could Climate Change Cause a Nuclear Meltdown?" *Government Accountability Project*, April 9, 2018. https://whistleblower.org/general/climate-science-watch/2018-04-09- as-fossil-fuels-melt-the-planet-could-climate-change-cause-a-nuclear-meltdown/.

President of Russia. 2022. "Joint Statement of the Russian Federation and the People's Republic of China on the International Relations Entering a New Era and the Global Sustainable Development." http://en.kremlin.ru/supplement/5770.

Proof (The Proof Is Out There). 2021. "Escape From Monkey Island." Season 1, Episode 2 of *The Proof Is Out There*. Premiered January 8, 2021. https://play.history.com/shows/the-proof-is-out-there/season- 1/episode-2.

Randall, Audrey. 2014. "Earthquakes and Nuclear Power Plants." *Science On a Sphere*, August 1, 2014. https://sos.noaa.gov/catalog/datasets/earthquakes-and-nuclear-power- plants/.

Ridge, Francis L. 2015 <2003>. "Nuclear Connection Project (NCP)." Accessed April 10, 2021. https://www.nicap.org/ncp/ncp-home.htm.

Rogoway, Tyler, and Joseph Trevithick. 2020a. "Mysterious Drone Incursions Have Occurred Over U.S. THAAD Anti-Ballistic Missile Battery In Guam." *The Drive*, September 14, 2020. https://www.thedrive.com/the-war-zone/36085/troubling-drone-   incursions-have-occurred-over-guams-thaad-anti-ballistic-missile- battery.

—. 2020b. "The Night A Mysterious Drone Swarm Descended On Palo Verde Nuclear Power Plant." *The Drive*, July 29, 2020. https://www.thedrive.com/the-war-zone/34800/the-night-a-drone- swarm-descended-on-palo-verde-nuclear-power-plant.

—. 2019. "The SR-71 Blackbird's Predecessor Created 'Plasma

Stealth' By Burning Cesium-Laced Fuel." *The Drive*, September 12, 2019. https://www.thedrive.com/the-war-zone/29787/ the-sr-71-blackbirds-predecessor-created-plasma-stealth-by-burning-cesium-laced-fuel.

Schwartz, David N. 2019. "How Scientific Chance and a Little Luck Helped Usher in the Nuclear Age." *Smithsonian Magazine*, January 17, 2019. https://www.smithsonianmag.com/ science-nature/enrico-fermi-scientific-chance-luck-nuclear-age-180971249/.

Singh, Kanishka. 2021. "U.S. President Biden signs $770 billion defense bill." *Reuters*, December 27, 2021. https://www.reuters.com/world/us/us-president-biden-signs-770-billion-defense-bill-2021-12-27/.

Smith, R. Jeffrey. 2021. "The US Nuclear Arsenal is Becoming More Destructive and Possibly More Risky." *The Center for Public Integrity*, October 29, 2021. https://publicintegrity.org/national-security/future-of-warfare/nuclear-weapon-arsenal-more-destructive-risky/.

Special access programs – congressional oversight. 2011. 10 U.S. Code §

119. https://www.govinfo.gov/content/pkg/USCODE-2011-title10/pdf/ USCODE-2011-title10-subtitleA-partI-chap2-sec119.pdf. [The date of this U.S. Code is 2011 and the law went into effect on January 3, 2012. Page numbers cited are from the PDF.]

Sturrock, Peter A. 1987. "An Analysis of the Condon Report on the Colorado UFO Project." *Journal of Scientific Exploration* 1, no. 1: 75-100.

—. 2004. "Time-Series Analysis of a Catalog of UFO Events: Evidence of a Local-Sidereal-Time Modulation." *Journal of Scientific Exploration* 18, no. 3: 399–419.

—. 1999. *The UFO Enigma: A New Review of the Physical Evidence.* New York: Warner Books, Inc.

Sturrock, Peter A., et al. 1998. "Physical Evidence Related to UFO Reports: The Proceedings of a Workshop Held at the Pocantico Conference Center, Tarrytown, New York,

September 29-October 4, 1997." *Journal of Scientific Exploration* 12, no. 2: 179–229. https://www.researchgate.net/publication/224791605_Physical_Eviden ce_Related_to_UFO_Reports_The_Proceedings_of_a_Worksh op_Held_at_the_Pocantico_Conference_Center_Tarrytown_N ew_York_Septe mber_29_-_October_4_1997.

Tingley, Brett. 2022. "Mysterious Drone Incursions Confirmed Over Sweden's Nuclear Facilities." *The Drive*, January 15, 2022. https://www.thedrive.com/the-war-zone/43901/mysterious-drone- incursions-confirmed-over-swedens-nuclear-facilities-this-weekend.

Toftenes, Terje, and Truls Toftenes, directors. 2010. *The Day Before Disclosure*. New Paradigm Films. https://topdocumentaryfilms.com/day-before-disclosure/?___ cf_chl_ f_tk=0JrpLcXL8fSqoMujNnB2jAIyyVR2pcABb8VH0JNf.cs-1642594230-0-gaNycGzNCKU.

Trevithick, Joseph. 2019. "Did Israel Just Conduct A Ballistic Missile Test From a Base On Its Mediterranean Coast?" *The Drive*, December 6, 2019. https://www.thedrive.com/the-war-zone/31358/did-israel-just-conduct- a-ballistic-missile-test-from-a-base-on-its-mediterranean-coast.

—. 2022. "Swedish Security Agency Declares A National Event As Drone Incursions Over Nuclear Sites Grow." *The Drive*, January 17, 2022. https://www.thedrive.com/the-war-zone/43905/swedish-security- agency-declares-a-national-event-as-drone-incursions-over-nuclear- sites-grow.

Tritten, Travis. 2021. "Air Force Veterans Who Are UFO True Believers Return to Newly Attentive Washington." *Military.com*, October 19, 2021. https://www.military.com/daily-news/2021/10/19/air-force-veterans- who-are-ufo-true-believers-return-newly-attentive-washington.html

UFOs & Nukes. n.d. "Documents." https://www.ufohastings.com/documents.

UFOs: The White House Files (UFOs). 2019. "UFOs: The White House Files." Season 1, Episode 1 of *UFOs: The White House Files. History Channel.* Premiered December 6, 2019.

University of Chicago Library. 2009. *Guide to the Association of Los Alamos Scientists Records 1945-1948.* The Hanna Holborn Gray Special Collections Research Center, University of Chicago Library. https://www.lib.uchicago.edu/e/scrc/findingaids/view.php?eadid=ICU. SPCL.ALAS.

USA War College (United States Army War College). 2019. Implications of Climate Change for the U.S. Army. file:///C:/Users/jensi/Desktop/implications-of-climate-change-for-us-army_army-war-college_2019.pdf.

U.S. DoD (U.S. Department of Defense). 2021. "DoD Announces the Establishment of the Airborne Object Identification and Management Synchronization Group (AOIMSG)." November 23, 2021. Washington, D.C.: U.S. Department of Defense. https://www.defense.gov/News/Releases/Release/Article/2853121/dod
-announces-the-establishment-of-the-airborne-object-identification- and-manag/.

—. 2020. "Establishment of Unidentified Aerial Phenomena Task Force." August 14, 2020. Washington, D.C.: U.S. Department of Defense. https://www.defense.gov/News/Releases/Release/Article/2314065/esta blishment-of-unidentified-aerial-phenomena-task-force/.

U.S. SASC (U.S. Senate Committee on Armed Services). 2021. Reed, Inhofe Praise Senate Passage of National Defense Authorization Act for Fiscal Year 2022. December 15, 2021. https://www.armed-services.senate.gov/press-releases/reed-inhofe- praise-senate-passage-of-national-defense-authorization-act-for-fiscal- year-2022.

Wikipedia. 2020. "A4W reactor." *Wikipedia: The Free Encyclopedia.* Accessed April 10, 2021. https://en.wikipedia.org/wiki/A4W_reactor.

—. 2021. "USS *Theodore Roosevelt* (CVN-71)." *Wikipedia: The Free Encyclopedia.* Accessed November 29, 2021. https://

en.wikipedia.org/wiki/USS_Theodore_Roosevelt_(CVN-71).

Woolsey, R. James. 2021. "Amb. R. James Woolsey on the Cold War, JFK Assassination & UFOs." *The Black Vault Originals*. Premiered April 2, 2021. https://www.youtube.com/watch?v=_vQF4QzUX9M&t=3s.

Woolsey, R. James, William R. Graham, Henry F. Cooper, Fritz Ermarth, and Peter Vincent Pry. 2021. "Iran Probably Already Has the Bomb. Here's What to Do about It." *National Review*, March 19, 2021. https://www.nationalreview.com/2021/03/iran-probably-already-has-   the-bomb-heres-what-to-do-about-it/.

—. 2016. "Underestimating Nuclear Missile Threats from North Korea and Iran." *National Review*, February 12, 20216. https://www.nationalreview.com/2016/02/iran-north-korea-nuclear/.

ABOUT THE AUTHOR

Jensine Andresen (Ph.D. Harvard University) holds a B.S.E. in Civil Engineering from Princeton University, where she also earned a Certificate from the School of Public and International Affairs. She completed an M.A. degree at Columbia University in Social Anthropology with a focus on China. She also earned A.M. (master's) and Ph.D. degrees at Harvard University from the Committee on the Study of Religion with a focus on Indo-Tibetan Buddhism. Dr. Andresen' translation of a renowned Sanskrit text and Tibetan annotations is forthcoming as *The Kālacakra Tantra: The Initiation Chapter with the Vimalaprabhā Commentary*.

Dr. Andresen served as a Visiting Assistant Professor at the University of Vermont, where she taught Science and Religion and world religions. She also was an Assistant Professor at Boston University in the interdisciplinary doctoral program on Science, Philosophy, and Religion. Dr. Andresen also held two academic appointments as a Visiting Scholar at Columbia University, and later she was appointed as an Officer of Research, Associate Research Scholar at Columbia.

Dr. Andresen is the author of *Extraterrestrial Ethics* (Ethics Press 2023). She edited *Religion in Mind: Cognitive Perspectives on Religious Belief, Ritual, and Experience* (Cambridge University Press 2001) and she is a co-editor of *Cognitive Models and Spiritual Maps: Interdisciplinary Explorations of Religious Experience* (Imprint Academic 2000) and of *Extraterrestrial Intelligence: Academic and Societal Implications* (Cambridge Scholars Publishing 2022). Dr. Andresen's recent chapter, "Two Elephants in the Room of Astrobiology," published in *Astrobiology: Science, Ethics, and Public Policy* (Wiley/Scrivener 2021) examines Unidentified Aerial Phenomena (UAP) in the context of the militarization and weaponization of space, which Dr. Andresen opposes. Her single-authored monograph, *Safe Space*, is forthcoming, as is her invited chapter "Which Science?" in *Astroanthropology: Science, Ethics, and Religion*, edited by Arvin Gouw, Brian Patrick Green, Junghyung Kim, and Ted Peters (forthcoming 2024). Her single-authored monograph, *Extraterrestrial Mind*, is in preparation.

Dr. Andresen has published multiple entries in *Encyclopedia of Science and Religion* (Macmillan, 2003), a chapter in *Fifty Years in Science and Religion: Ian G. Barbour and his Legacy* (Ashgate, 2004), and articles in many peer- reviewed journals and other publications, including *The International Journal for the Psychology of Religion, Harvard Theological Review, Dreaming, The American Society of International Law, Proceedings of the 96th Annual Meeting, March 13-16, 2002, Washington, DC: The Legalization of International Relations/The Internationalize of Legal Relations, Journal of Cultural Diversity, Journal of Sleep Research, The Journal of Religion, Zygon: Journal of Religion and Science, Religion and Education, Isis (Supplement, Catching up with the Vision: Essays on the Occasion of the 75th Anniversary of the Founding of the History of Society Society)*, and Boston University's *Focus*. Dr. Andresen is one of the co-authors of *Report on Ecumenical Faith and Genetics Working Group*, which was written as part of work with the Episcopal Diocese of Massachusetts Faith and Genetics Working Group. She also created a six-part videotape series, *Bioethics and Society: Scientific, Ethical, Legal, and Religious Perspectives on Genetic Technologies.*

In addition to her work in academia, Dr. Andresen has held various positions in finance, business, and government. Dr. Andresen may be reached by email at: jensine@alumni.princeton.edu.

www.ingramcontent.com/pod-product-compliance
Lightning Source LLC
Chambersburg PA
CBHW062244290526
45794CB00006B/2392